Water, Sanitation, Hygiene, and Nutrition in Bangladesh

A WORLD BANK STUDY

Water, Sanitation, Hygiene, and Nutrition in Bangladesh

Can Building Toilets Affect Children's Growth?

Iffat Mahmud and Nkosinathi Mbuya

WORLD BANK GROUP

Contents

Tables

Acknowledgments

The authors of the report would like to thank Mduduzi Mbuya (Sanitation Hygiene, Infant Nutrition Efficacy Study, ZVITAMBO) for his extensive technical inputs to the report and for the modified "Child-friendly" WASH Framework. We also thank Rokeya Ahmed (Water and Sanitation Specialist, the World Bank) for her contribution and extensive support in preparing the report. The authors are grateful to Albertus Voetberg (Interim Practice Manager of Health, Nutrition, and Population Global Practice of the World Bank), under whose oversight this analytical work was conducted.

The peer reviewers of the draft report from the World Bank were Meera Shekhar (Lead Health Specialist), Susanna Smets (Senior Water and Sanitation Specialist), and Dinesh Nair (Senior Health Specialist). The authors have greatly benefited from the helpful comments provided by the peer reviewers. Detailed comments were also provided by Emily Christensen Rand and the authors are thankful to her. The authors appreciated the valuable input provided by Iffath Sharif (Program Leader) on an earlier version of the report. The authors would also like to acknowledge the contributions of Arshee Rahman, Fariha Nehreen Mirza, and Saadat Chowdhury.

Finally, the authors express their gratitude to Johannes Zutt (Country Director of the World Bank), who chaired an internal review meeting of the draft to finalize its contents and to discuss its multisector implications. Michael Alwan edited the report.

Executive Summary

Since the 1960s, it has been known that poor water and sanitation causes diarrhea, which consequently compromises child growth and leads to undernutrition. Ample evidence shows that poor water and sanitation causes diarrhea, but there is a growing body of knowledge discussing the magnitude of the impact of diarrhea on undernutrition. A recent hypothesis by Humphrey (2009), for example, states that the predominant impact of contaminated water and poor sanitation on undernutrition is via *tropical/environmental enteropathy* (triggered by exposure to fecal matter) rather than mediated by diarrhea. This new hypothesis has generated much debate, especially in the South Asia region, on the contribution of water and sanitation to the South Asian Nutrition Enigma. The region is characterized by unusually high rates of child undernutrition relative to its income level, as well as a slow reduction in undernutrition. Practitioners have struggled to decipher the reasons behind this "anomaly."

This report provides a systematic review of the evidence to date, both published and grey literature, on the relationship between water and sanitation and nutrition. We also examine the potential impact of improved water, sanitation, and hygiene (WASH) on undernutrition. This is the first report that undertakes a thorough review and discussion of WASH and nutrition in Bangladesh. The report is meant to serve two purposes. First, it synthesizes the results/evidence evolving on the pathway of WASH and undernutrition for use by practitioners working in the nutrition and water and sanitation sectors to stimulate technical discussions and effective collaboration among stakeholders. Second, this report serves as an advocacy tool, primarily for policy makers, to assist them in formulating a multisectoral approach to tackling the undernutrition problem.

Bangladesh has achieved remarkable progress in overall health outcomes (particularly in reducing fertility and child and maternal mortality), but not as much in nutrition. Impressive progress has also been made in vitamin A supplementation (which may have contributed to the reductions in child mortality), salt iodization (which has significantly reduced goiter rates), and iron supplementation. However, commensurate gains in nutritional outcomes have not been witnessed. Although undernutrition in Bangladesh has declined gradually since the 1990s, the prevalence remains high. According to the Utilization of Essential Services Delivery Survey (UESD), 38.7 percent of children under five years of age are stunted (short for their age), and 35 percent are underweight (low weight for age), in 2013 (NIPORT 2013). As per the classification of the World Health Organization (WHO), there is "very high prevalence" of underweight in Bangladesh, a rate that is higher than most Sub-Saharan Africa. Moreover, under-nutrition does not only affect the poor people of Bangladesh. Undernutrition rates are also relatively high among the wealthy: 24 percent of children under five years of age were underweight in the richest quintile in 2013.

The undernutrition problem has primarily been dealt with through health sector interventions in Bangladesh. These have successfully increased nutrition-related knowledge and attitudes, but have had limited impact on nutritional outcomes. Up to 2011, the Ministry of Health and Family Welfare (MOHFW) had imple-mented direct nutrition interventions through a community-based approach covering 172 upazilas in phases (out of the total of 488 upazilas in Bangladesh). The MOHFW contracted nongovernmental organizations (NGOs) to reach out to the communities and deliver a package of services. An evaluation of these interventions revealed that although these have been successful in improving caregiving practices (such as health and nutrition–related knowl-edge and attitudes, as well as some key feeding practices), these improvements did not reduce poor nutritional outcomes such as underweight or stunting (Mbuya and Ahsan 2013). In 2011, the MOHFW initiated the provision of the basic nutrition services through the various tiers of the public health facilities nationwide. A mid-term assessment of this new modality indicated that the overall effort is an ambitious, but valuable, approach to examining how best to support nutrition actions through an existing health system with diverse plat-forms (World Bank 2014). The assessment highlighted the need to strengthen the intervention through better coordination and prioritization of a set of key activities.

The causes of undernutrition are multifactorial and calls for both "nutrition-specific" as well as "nutrition-sensitive" actions from multiple sectors, not just health. Substantial global evidence shows that direct actions to address the immediate determinants of undernutrition ("nutrition-specific") can be further enhanced by actions addressing the more underlying determinants ("nutrition-sensitive"). These "nutrition-sensitive" actions are in the domains of ministries other than health, hence the need for a more comprehensive, multisectoral approach to address maternal and child undernutrition (Gillespie et al. 2013). Although the health sector in particular must lead the effort on nutrition, it is clear that other key sectors need to ensure that their own policies and programs are "nutrition-sensitive." They need to provide the requisite support to deliver on nutrition's potential for Bangladesh (World Bank, DFID, Government of Japan, and Rapid Social Response 2013).

Improved WASH interventions are necessary for reducing undernutrition but not sufficient to create a dent in the undernutrition problem. Adequacy of food, health care, and WASH are all critical for reducing undernutrition. To fully realize the impact of WASH interventions, multisectoral actions are needed. For example, it is important to design the programs so that they address a full spectrum of WASH-related issues. These include clean water, proper sanitation facilities, and reduction of fecal matter (both human and animal) in the environment (including soil and children's play areas); availability of water and soap for handwashing; and behavioral issues, such as instilling the habit of handwashing with soap at critical times (after using the toilet, before preparation of food, after cleaning babies, before eating, and so forth). Furthermore, WASH efficacy depends on combined efforts on three fronts: improving the quality and quantity of food; ensuring adequate childcare practices (ensuring that children are immunized and pregnant women can access antenatal and postnatal care services); and improving WASH interventions (Newman 2013), with emphasis particularly on the "H" for hygiene. Inasmuch as building a toilet and reducing open defecation will not translate into the growth of a child, food alone might not be adequately absorbed and utilized—making the various dimensions necessary but not necessarily, sufficient.

The diarrhea-undernutrition hypothesis postulated that diarrhea is both a cause and effect of undernutrition. Recent evidence, however, contends that the effect of diarrhea on undernutrition is not as significant as previously thought. Children with diarrhea have depressed appetite and are less able to absorb the nutrients from their food, and undernourished children are more susceptible to diarrhea when exposed to fecal bacteria from their environment. This synergistic relationship, while still valid, appears not to result in long-term undernutrition (stunting). In 2013, a meta-analysis in the Cochrane review found that the WASH interventions have resulted in only moderate increases in weight and height and have not had a significant effect on undernutrition (Dangour et al. 2013). This weak linkage between diarrhea and undernutrition is very relevant in interpreting the successful management of diarrhea in Bangladesh. An extensive oral rehydration program has reduced the prevalence of diarrhea among children under five years of age. However, this has not translated to a comparable effect on nutritional (anthropometric) outcomes.

Humphrey (2009) hypothesized that the predominant causal pathway from poor sanitation and hygiene to undernutrition is tropical/environmental enteropathy, not diarrhea. Both the diarrheal and the tropical/environmental enteropathy hypotheses are premised upon fecal-oral contamination. However, it is the biological "response" to the fecal-oral contamination that is different. Diarrhea is a clinical condition and results in loss of appetite and nutrients, whereas tropical/environmental enteropathy is a physiological condition without signs/symptoms (that is, subclinical). It is characterized by physiological and anatomical changes to the structure of the small intestine that affect a child's ability to both absorb and utilize nutrients. In a seminal *Lancet* publication, Humphrey (2009) hypothesized that infants and young children living in conditions of poor sanitation and hygiene have chronic exposure to large quantities of fecal bacteria, which results in a subclinical disorder of the small intestine known as tropical/environmental enteropathy.

Tropical/environmental enteropathy is characterized by decreased villous height and increased permeability of the intestinal tract. Villi are small fingerlike projections on the intestinal wall that provide a large surface area for absorption of nutrients. The reduction in the height of villi in small intestine reduces the total area of the small intestine, and, therefore, the absorption of nutrients is lowered. At the same time, an increase in porosity of the intestinal tract reduces the ability of the body to prevent pathogens from crossing the intestinal barrier, which triggers the response of the immune system and diverts nutrients for use toward defending against pathogens rather than toward supporting the normal growth of the child.

In the water and sanitation sector in Bangladesh, progress has been made in coverage of water and sanitation facilities, but not in hygiene promotion. The qualities of water and sanitation facilities also need improvements. The key achievement in sanitation has been the shift from open defecation to "fixed point defecation." Open defecation has been reduced to 3 percent of the population in 2012 from 42 percent in 2003. However, only 57 percent of the population uses an "improved" sanitation facility (WHO/UNICEF 2013) and one-third of the households share latrines. In water, the transitioning from traditional sources (such as ponds and canals) to piped or improved sources (mostly tubewells and piped water) has been considered as a significant achievement. Further improvements in quality of water should be prioritized, as only 10 percent of the population has access to water piped to the premises. To ensure sustainability of the public goods, the government will need to increase its share of financing and ensure routine monitoring (at present, 35 percent of the total funding available to the water and sanitation sector is from the government).

Reducing open defecation cannot be considered as "mission accomplished." The key issue is to prevent exposure to fecal matter (human and animal) in order to hinder initiation of tropical/environmental enteropathy. Building toilets and providing reliable sources of water supply, therefore, will not yield much unless the quality of water and sanitation facilities is improved and hygiene practices are promoted to reduce fecal-oral contamination. Renewed political stewardship is required for promotion of improvements in water and sanitation facilities.

Hygiene remains the weakest link. According to the Bangladesh National Baseline Hygiene Survey 2014, although more than two-thirds of the households had a location near the toilet for postdefecation handwashing, only 40 percent had water and soap available. During handwashing demonstrations, only 13 percent of children aged three to five years of age and 57 percent of mothers/female caregivers washed both hands with soap. The Department of Public Health and Engineering (DPHE), the main agency working in the water and sanitation sector, focuses on infrastructure development and does not have the capacity, neither the comparative advantage of implementing behavior change communication (BCC) activities.

The Government of Bangladesh (GOB) has formulated a set of comprehensive policies and strategies in the water and sanitation sector (four legislative acts, two national policies, and five national strategies). However, translation of these policies and strategies into action appears to be a challenge. There are bottlenecks in implementing some of these strategies as this requires a well-coordinated set of actions by multiple ministries. There are not enough incentives and scope for the individual ministries to work beyond their domains.

Against this backdrop, the following recommendations are made, which may be considered by the GOB. The first two sets of recommendations are specific for the water and sanitation and the health sectors, while the third set of recommendations is overarching and multisectoral. A framework is also presented in table 4.6 that provides an overview of the key factors and potential areas of intervention in reducing contamination of fecal matter. These include reducing fecal load in the living environment, reducing fecal transmission via unclean hands, improving quality of drinking water, and avoiding ingestion of chicken feces by children while playing.

Recommendations

Make Water and Sanitation Activities More "Nutrition-Sensitive"

1. **Improve quality of water and sanitation facilities.**
   It is critical to improve the quality of water (at the source, in storage, and at the point of consumption)—and sanitation facilities to limit transmission of infection. There is also a need to ensure that households that have a piped water supply also have water that is safe for drinking. Awareness campaigns along with emotional/social drivers can be effective in meeting these needs.

2. **Strengthen implementation of hygiene-related activities.**
   Hygiene remains the weakest link in the water and sanitation sector. At the strategic level, the 2014 draft National Water Supply and Sanitation Strategy adequately addresses this issue. It is now, therefore, critical to finalize the draft 2014 Strategy and implement the action plan. The GOB will need to monitor progress of the implementation of the action plan through a high-level intersectoral committee. Particular emphasis should be placed on increasing the availability of handwashing stations and ensuring that these are used.

Improve the Nutrition Activities of the Health Sector

1. **Strengthen the effectiveness of the National Nutrition Services (NNS).**
   The MOHFW should define and prioritize a critical set of activities for improving undernutrition, particularly improved hygiene practices. As the recent NNS assessment indicates that the current delivery platform is not being effective, alternative service delivery mechanisms will need to be explored to extend outreach and achieve greater targeted coverage. NNS, due to its modality of service delivery through public health facilities, is targeted

toward mostly the poorer and disadvantaged population. The MOHFW, therefore, will continue to overlook large segments of the population where undernutrition rates are high. The MOHFW may actively consider engaging the media and the private sector for the required BCC as well as promoting handwashing through the health sector interventions.

2. **Consider the preventative aspects of nutrition, rather than just treatment.**
At present, under NNS, small corners for integrated management of childhood illnesses and nutrition ("IMCI&N corners") are being set up at the health facilities. This modality has the disadvantage of only covering sick children by the nutrition services. The MOHFW needs to transition from the IMCI&N corners to investing more deeply in an alternative and predominantly outreach-based platform for delivering core services to households and children. These might be called "well-child spots" and be located near or at the existing health facilities (World Bank 2014).

Enact a Multisectoral Response to Undernutrition

1. **Strengthen the health sector response, but also build a nonhealth, multisectoral response for addressing undernutrition.**
The determinants of undernutrition are multisectoral, yet attempts to implement multisectoral programs have proved largely unsuccessful. Multisectoral nutrition planning agencies have been stymied by the limited control they have over different sectors' resource allocation processes, while sectorally defined priorities have hindered collaboration between sectors. A more realistic response is to "plan multisectorally, implement sectorally" (Maxwell and Conway 2000). Operationally, this involves identifying interventions within sectors that have the potential to significantly improve nutrition and mobilizing resources specific to that sector (World Bank, DFID, Government of Japan, and Rapid Social Response 2013).

2. **Align efforts of the various sectors with the overall goal of reducing undernutrition.**
Individual efforts by MOHFW and other ministries have the desired impact on undernutrition rates. The relevant sectors—health, nutrition, and population (HNP); water and sanitation; education; local government; agriculture—need to integrate their efforts to attain the broader national goal of improving nutritional outcomes. To enable this, alleviating undernutrition must remain a high-level policy priority. Promoting interventions with cross-sectoral benefits will be useful.

Abbreviations

BCC	Behavior Change Communication
BDHS	Bangladesh Health and Demographic Survey
BDWSSP	Dhaka Water Supply and Sanitation Project
BETV-SAM	Bangladesh Environmental Technology Verification-Support to Arsenic Mitigation
BINP	Bangladesh Integrated Nutrition Project
BNNC	Bangladesh National Nutrition Council
BRWSSP	Bangladesh Rural Water Supply and Sanitation Project
BWDB	Bangladesh Water Development Board
CLTS	Community-Led Total Sanitation
CNP	Community Nutrition Provider
CWASA	Chittagong Water and Sewerage Authority
CWSSP	Chittagong Water Supply Improvement and Sanitation Project
DP	Development Partner
DPHE	Department of Public Health and Engineering
DWASA	Dhaka Water Supply and Sewerage Authority
ETV-AM	Environmental Technology Verification-Arsenic Mitigation
FAO	Food and Agriculture Organization
FSNSP	Food Security and Nutritional Surveillance Survey
GOB	Government of Bangladesh
HKI	Helen Keller International
HNP	Health, Nutrition, and Population
HNPSP	Health, Nutrition, and Population Sector Programme
HPNSDP	Health, Population, and Nutrition Sector Development Programme
IFAD	International Fund for Agricultural Development
IFPRI	International Food Policy Research Institute
IMCI&N	integrated management of childhood illnesses and nutrition
IPHN	Institute of Public Health and Nutrition
JMP	Joint Monitoring Programme

LGD	Local Government Department
LGED	Local Government Engineering Department
LGI	Local Government Institution
MDG	Millennium Development Goal
MICS	Multiple Indicator Cluster Survey
MOHFW	Ministry of Health and Family Welfare
MOLGRD&C	Ministry of Local Government Rural Development and Cooperatives
MOPME	Ministry of Primary and Mass Education
MOSHE	Ministry of Secondary and Higher Education
NFWSS	National Forum for Water Supply and Sanitation
NGO	nongovernmental organization
NNP	National Nutrition Project
NNS	National Nutrition Services
NPAN	National Plan of Action for Nutrition
NSAPR	National Strategy for Accelerated Poverty Reduction
PA	Project Aid
PPRC	Power and Participation Research Centre
RPA	Reimbursable Project Aid
SDP	Sector Development Plan for the Water Supply and Sanitation Sector
SDR	Special Drawing Rights
UESD	Utilization of Essential Service Delivery
UN	United Nations
UNDP	United Nations Development Programme
UNICEF	The United Nations Children's Fund
WARPO	Water Resources Planning Organization
USAID	U.S. Agency for International Development
WASA	Water Supply and Sewerage Authority
WASH	Water Sanitation and Hygiene
WATSAN	Water Supply and Sanitation Committees
WFP	World Food Programme
WHO	World Health Organization

CHAPTER 1

Introduction

Background and Rationale

Bangladesh faces an unfinished agenda with regard to nutrition. Between 2004 and 2013, among children under five years of age, underweight rates declined from 43 percent to 35 percent and stunting rates declined from 51 percent to 39 percent (NIPORT 2013). Progress with regards to nutritional outcomes is less than satisfactory and child undernutrition rates in Bangladesh remain among the highest in the world.

Maternal undernutrition, a key determinant of infant and young child undernutrition, remains intractable despite efforts to improve the nutritional status of pregnant women. For example, iron deficiency anemia affects nearly half of all Bangladeshi pregnant and lactating women and is directly related to low birth weight, which affects a large proportion of all newborns. The high levels of maternal and child malnutrition are of grave concern, given that malnutrition between conception and two years of age cause irreversible damage to a child's health, growth, and cognitive development. Malnutrition also contributes to higher child morbidity and mortality, lower intelligence quotient, lower school achievement; reduced adult productivity, and lower earnings. It has been estimated that undernutrition costs Bangladesh more than Tk 70 billion (or US$10 billion) in lost productivity every year, and even more in health care costs (FAO, WFP, and IFAD 2012).

To date, public sector investments to address undernutrition in Bangladesh have had very little impact because they have been limited in both scope and scale. Undernutrition is a multidimensional problem requiring interventions that cut across sectoral boundaries. According to a framework developed by United Nations Children's Fund (UNICEF) in the 1990s, now widely accepted and globally used, undernutrition is an outcome of immediate, underlying, and basic causes. At the immediate level, nutritional status is determined by the availability of nutrients to the body to meet its requirements and the status of health, while the underlying and basic causes include food security (access, availability, and utilization of food), maternal and child caring practices, water and sanitation, and personal hygiene. These determinants are heavily influenced by the social status of women, as well as institutional/organizational, political and ideological, economic, and environmental constraints. As such, sustained improvements in child nutritional outcomes can be achieved not only through improved food security but also through changes in behavior, knowledge, and attitude within the household regarding maternal and childcare, appropriate feeding practices, and health care. Such changes require broader interventions that cut across multiple sectors, including food and agriculture, water and sanitation, education, and health. The potential for improving nutrition through interventions in these sectors has not been fully exploited in the context of Bangladesh.

Nutrition interventions in Bangladesh have largely been implemented through the health sector. These interventions tackle undernutrition by addressing behavioral issues surrounding caring and feeding practices, providing multi-micronutrients, therapeutic/supplemental foods, and improving access to health care. However, health sector interventions have not been undertaken in a multisectoral approach or context.

There is now mounting global evidence from diverse sources—including biological, epidemiological, and econometric analyses—of a strong linkage between poor sanitation and hygiene and child undernutrition. Over the last five or so decades, diarrhea has been implicated as the most significant intermediate factor in the causal pathway from poor sanitation to undernutrition. However, more recent hypotheses and analyses suggest that the impact of diarrhea on long-term growth (stunting) may not be as substantial as previously postulated. The main reason is that growth velocity can be faster than average-for-age between illness episodes, resulting in catch-up growth. A recent hypothesis suggests that "tropical/environmental enteropathy" is a major contributor of undernutrition in the poor sanitation–undernutrition causal pathway.

Objectives

The objectives of this report are to (i) examine the pathways of improving nutrition through interventions in the water and sanitation sector; (ii) explore ways that could potentially improve nutritional outcomes through interventions in water and sanitation; and (iii) explore how these can be integrated in a better-coordinated multisectoral approach to address undernutrition. Specifically, the report reviews the evidence (both published and grey literature), policies, and programs related to water and sanitation that are potentially influential for nutrition. Also, it undertakes a comparison of what could be done and what is actually being carried out in the water and sanitation sector in Bangladesh. The report, hence, proposes an outline of action for future consideration and support from the Government of Bangladesh (GOB) and its development partners (DPs).

Note that this report is exploratory in nature and does not constitute an exhaustive review of the impact of interventions in the water and sanitation sector on undernutrition. It is based entirely on secondary data and desk reviews of the published and grey literature and does not use primary data to evaluate interventions and intervention approaches.

Structure of the Report

Chapter 1 (Introduction) provides the background, rationale, and objectives of this work. Chapter 2 assesses the status of undernutrition in Bangladesh, provides a brief history of policies and programs to address undernutrition in the country, and lays out the case for a better-coordinated multisectoral response to undernutrition. Chapter 3 reviews the pathways through which water and sanitation outcomes can impact undernutrition. This chapter explains the evolution of the theory and evidence of the contribution of water and sanitation to undernutrition as well as the new hypothesis of tropical/environmental enteropathy. Chapter 4 provides an overview of the policies, strategies, and interventions in the water and sanitation sector in Bangladesh. A theoretical framework is presented in this chapter that will assist planners and implementers in devising effective interventions in the water and sanitation sector that can have the maximum impact on undernutrition. Recommendations for sector-specific as well as multisectoral actions are provided in chapter 5.

CHAPTER 2

Policy and Programmatic Responses to Undernutrition in Bangladesh: Why Coordinated Multisectoral Actions Are Needed

Key Messages

- Child undernutrition rates in Bangladesh remain among the highest in the world, despite the impressive progress with respect to the health-related Millennium Development Goals (MDGs).
- Underweight and wasting rates in Bangladesh are at a "very high" level of prevalence by World Health Organization (WHO) standard. In 2013, 38.7 percent of children under five years of age were short for their age (stunted), 18 percent had low weight for their height (wasted), and 35 percent had low weight for age (underweight).
- The underweight rate in Bangladesh (35 percent in 2013) is higher than that of Sub-Saharan Africa (30 percent in 2012).
- The stunting rate declined in Bangladesh from 51 percent to 39 percent and the underweight rate declined from 43 percent to 35 percent between 2004 and 2013. The wasting rate, however, increased from 15 percent to 18 percent over the same period of time.
- Poverty does not appear to be a key cause of undernutrition—in the richest income quintile in 2013, 24 percent was underweight. Underweight rates increased by three percentage points between 2011 and 2013 in the richest income quintile, while the rate declined by six percentage points in the poorest income quintile over the two-year period.
- Undernutrition is caused by multiple factors. To date, the problem has largely been addressed through health sector interventions. There is now a need to prioritize nutrition in other sectors as well to make sustained improvements in nutritional outcomes.

- "Nutrition-specific" interventions by the health sector can be enhanced by "nutrition-sensitive" actions that lie in the domains of other sectors (including agriculture, water and sanitation, and education).

Levels and Trends in Undernutrition in Bangladesh

Bangladesh has made significant strides in economic development over the years despite political turbulence and vulnerability to natural disasters. The country has averaged a steady economic growth rate of 5.8 percent annually during the past decade; maintained relatively low inflation; and has had fairly stable domestic debt, interest, and exchange rates. Bangladesh has made laudable progress on many aspects of human development (World Bank 2012). In education, Bangladesh has experienced impressive gains in improving access to education, reaching the Millennium Development Goal (MDG) gender parity at the primary and secondary levels. These are remarkable feats, given the enormous challenge that the country faced just a decade ago, and also in comparison to several other countries in the region. In the health, nutrition, and population (HNP) sector, impressive declines in infant and child mortality rates and maternal mortality ratio have put the country on track to meet MDGs 4 and 5 respectively. For the reduction in child mortality rate, Bangladesh was awarded the United Nations MDG Award 2010. All these factors, plus increased female job opportunities, have contributed to reducing the fertility rate by 60 percent since the 1970s (one of the fastest declines in the world). Commensurate improvements in nutrition outcomes have not, however, been witnessed by the country as indicated in the following section. The 35 percent underweight rate in 2013 (30 percent in 2012) among children under five years of age in Bangladesh (NIPORT 2013) is higher than that most of Sub-Saharan Africa, despite the latter's higher poverty rates (United Nations 2014). Extreme poverty in Sub-Saharan African was 48 percent in 2012 (United Nations 2014), while in Bangladesh it was 17.6 percent in 2010 (World Bank 2013a).

Child Undernutrition

The underweight and wasting rates in Bangladesh are at a "very high" level according to the standards of the World Health Organization (WHO) (table 2.1 and figure 2.1). Undernutrition in Bangladesh has declined gradually since the 1990s, but prevalence remains high: In 2013, 38.7 percent of children under five years of age were stunted (short for their age), 18 percent were wasted (low weight for height), and 35 percent were underweight (low weight for age). Despite these reductions in stunting and underweight, rates of wasting increased gradually between 2004 and 2014 (figure 2.1).

    Stunting (low height for age) is an indicator of chronic undernutrition. Its multifaceted causes include poor infant and young child feeding practices, frequent infections, poor access to food and health care, inadequate sanitation and handwashing practices, poor maternal education, child marriage, early first birth, and the degraded status of girls and women in the family and in society.

Table 2.1 Public Health Significance of Undernutrition in Bangladesh, 2013

	WHO classification (prevalence %)			
	Low	Medium	High	Very High
Stunting	<20	20–29	30–39	≥40
Underweight	<10	10–19	20–29	>30
Wasting	<5	5–9	10–14	>15
	Bangladesh (prevalence %)			
Stunting		38.7		
Underweight				35.1
Wasting				18.1

Sources: WHO and authors' calculations from NIPORT 2013.
Note: WHO = World Health Organization.

Figure 2.1 Undernutrition Trends in Bangladesh, 2004–13 (%)

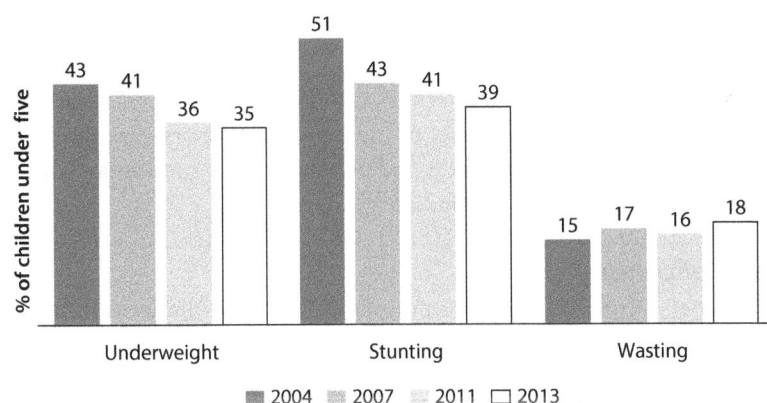

Sources: Data from NIPORT et al. 2013 and NIPORT 2013.

Stunting during the first two years of life has been associated with negative and long-lasting health, cognitive/schooling, and economic consequences.

A striking finding of the Bangladesh Demographic and Health Survey (BDHS) 2011 data (NIPORT et al. 2013), and confirmed elsewhere (PPRC and UNDP 2012), is that overall indicators of economic growth and greater household wealth are not strongly related to improved nutrition. Undernutrition does not seem to be a phenomenon found only among the poor people of Bangladesh—with one in three children (35 percent) under five years of age underweight and two in five children (39 percent) stunted even in the highest household wealth quintile in 2013.

Rates of undernutrition are quite high among the richer segment of the population. Moreover, a comparison of the data on undernutrition between 2011 and 2013 reveals that the rates of underweight and wasting have increased among the households of the highest wealth quintiles, while stunting has remained the same (figure 2.2). Over the same period of time, the rates of underweight and stunting have shown a modest decline among households of the lowest wealth quintiles.

Figure 2.2   Undernutrition Trends in Bangladesh by Economic Status, 2011–13 (%)

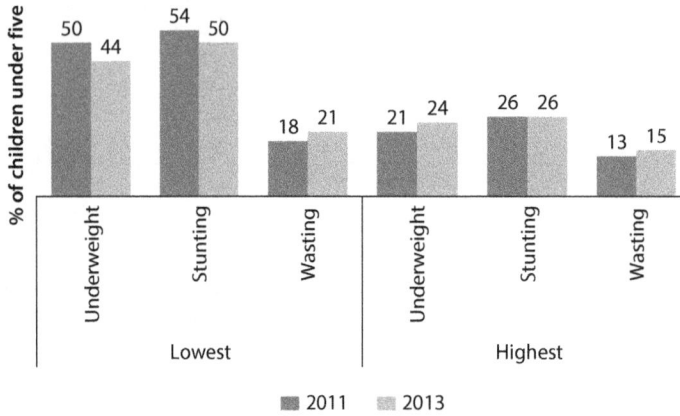

Sources: Data from NIPORT et al. 2013 and NIPORT 2013

One of the reasons for this could be the lack of knowledge or awareness regarding undernutrition. Also, in Bangladesh, most of the national programs are targeted to the people of the lower economic status and particularly the rural areas, which could be a reason for this unexpected finding.

Micronutrient Deficiencies

Bangladesh has successfully reduced the prevalence of night blindness induced by vitamin A deficiency among children. Vitamin A deficiency has been identified as a public health problem since the 1960s and the single most important preventable cause of night blindness in children. In particular, subclinical vitamin A deficiency among preschool children was classified as a problem of public health significance. In 2011–12, the prevalence of subclinical vitamin A deficiency was 20.5 percent in children of preschool age. According to the WHO classification, Bangladesh has mild vitamin A deficiency (cut-off value of < 1.05 micromoles/liter). High levels of vitamin A deficiency are associated with increased risk of mortality in children. Over the last 25 years, the Government of Bangladesh (GOB) initiated a vitamin A supplementation program targeted at children aged 6–59 months. The success of the supplementation program has been sustained with high coverage rates—in 2013, 80 percent of the target population, with a gap of 10 percentage points between top and bottom socio-economic quintiles. This has kept vitamin A deficiencies at a relatively low level. The coverage of the vitamin A supplementation program has increased over time and there are few disparities across income groups (figure 2.3).

During pregnancy and early childhood period, insufficient iodine causes varying degrees of irreversible brain damage. The problem of iodine deficiency in Bangladesh is classified as "mild" as the GOB has successfully promoted the production and consumption of iodized salt. The prevalence of goiter, which

Figure 2.3   Vitamin A Supplementation in Bangladesh by Wealth Quintile, 2011–13 (%)

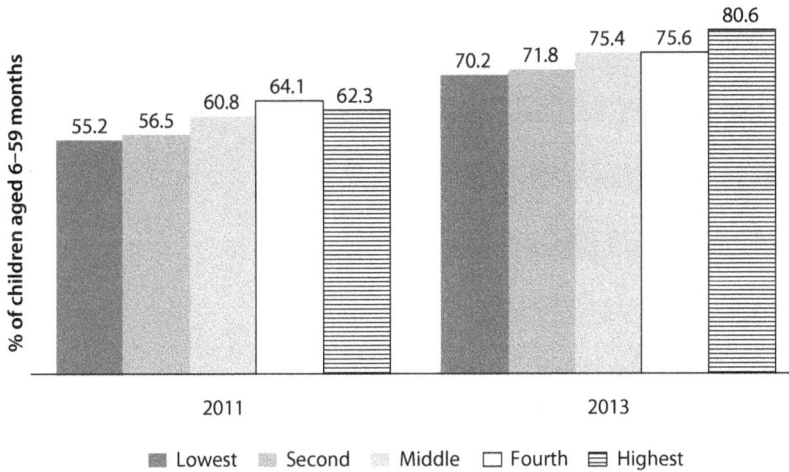

Sources: Data from NIPORT et al. 2013 and NIPORT 2013.

is the most visible form of iodine deficiency, consequently decreased from 49.9 percent in 1993 to 6.2 percent in 2004 among school children and from 55.6 percent to 11.7 percent among women over the same period (Yusuf et al. 2008).

Bangladesh, however, has not been able to successfully reduce anemia as much. As per 2011 BDHS data, almost half (51 percent) of children aged 6–59 months suffered from some level of anemia (hemoglobin of less than 11 grams per deciliter; Hb <11.0 g/dl)—29 percent of children had mild anemia (Hb 10.0–10.9 g/dl), and 21 percent had moderate anemia (Hb 7.0–9.9 g/dl). From the 2011 data, it appears that the prevalence of anemia peaks at 9–17 months (76–79 percent). The rates of anemia among children did not vary much by mother's education or economic status of the household. The iron folate supplementation program was equitable in distribution among the wealth quintiles (figure 2.4).

Evolution of Nutrition Policies and Interventions in Bangladesh

National Agencies and Policies to Address Uundernutrition

In one of the earliest attempts by the GOB to address undernutrition in a comprehensive manner, the Bangladesh National Nutrition Council (BNNC) was established in 1975 by an order of the president. This high-level agency was made responsible for the overall coordination of nutrition policy. Its tasks included the formulation of the National Food and Nutrition Policy, coordination of nutrition programs across different ministries and institutes, monitoring and evaluation of nutrition programs, and the preparation of a national plan for

Figure 2.4   Iron Supplementation in Bangladesh by Wealth Quintile, 2011 (%)

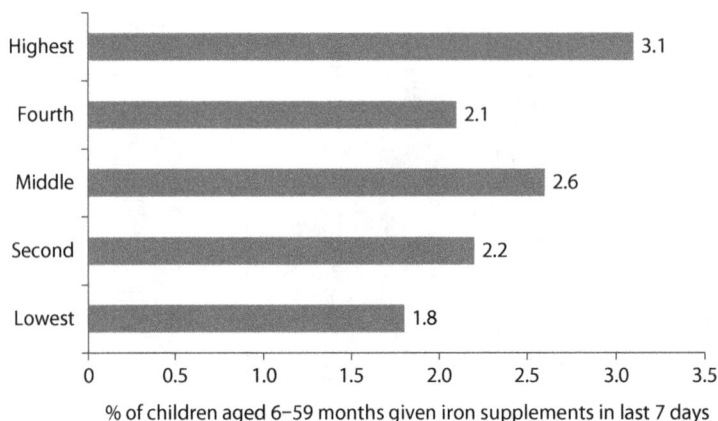

% of children aged 6–59 months given iron supplements in last 7 days

Source: Data from NIPORT et al. 2013.

nutrition. However, since its formation, the BNNC has met only twice and has not been functional for over a decade now.

Nutrition is highlighted as a priority area of intervention for the GOB in most national policy documents, particularly the National Strategy for Accelerated Poverty Reduction II (NSAPR, 2009), National Health Policy (2011), the National Strategic Plan for the Health, Population and Nutrition Sector Development Programme (2011–16), and the Perspective Plan (2010–21). Earlier in 1997, Bangladesh produced a National Plan of Action for Nutrition (NPAN), inspired by the International Conference on Nutrition five years earlier. The primary objective of NPAN was to improve the nutritional status of the people of Bangladesh so that undernutrition would no longer be a public health problem by 2010. NPAN has been implemented on an ad hoc basis over the years and as a result has remained largely ineffective.

The NSAPR is the current overarching public policy for combating poverty in all its dimensions, including undernutrition. It identifies specific avenues through which poverty reduction will be achieved. Nutrition has been included primarily under the heading of health and sparingly throughout the document. The National Health Policy formulated in 2011 aims to reduce the prevalence of undernutrition, especially among the children and mothers, and undertake effective and integrated programs to improve their nutritional status. The Perspective Plan (2010–2021), which provides a roadmap for accelerated growth, offers "broad approaches for eradication of poverty, inequality and human deprivation." For improving nutrition, the Perspective Plan identifies the role of nonhealth sectors and highlights key strategies, including better education in health and hygiene, reduction in the incidence of diarrhea, use of pure drinking water, diversification of agricultural productions, improved knowledge of balanced diet and nutrition, and so forth.

Policies and Interventions to Address Nutrition in the Health Sector

The undernutrition problem in Bangladesh has largely been viewed through a "health sector lens." The Bangladesh Integrated Nutrition Project (BINP, 1995–2002), with a budget of US$67 million, was the first national program to tackle undernutrition in the country. The project was implemented in 40 rural upazilas between 1995 and 2000, and was expanded to a further 21 upazilas by 2002 covering approximately 16 percent of the rural population (in the 61 upazilas). In 2003, BINP was succeeded by the National Nutrition Project (NNP), which expanded coverage to a total of 110 upazilas. NNP was funded under a separate World Bank credit worth US$124 million. In 2006, NNP was merged into the Health, Nutrition and Population Sector Programme (HNPSP) and was implemented in 172 upazilas covering 34 percent of the population. HNPSP, a US$4 billion program from 2006 to 2011, was cofinanced by the GOB and development partners (DPs) and implemented by Ministry of Health and Family Welfare (MOHFW) using a sectorwide approach.

The scope of nutrition services provided through BINP, NNP, and HNPSP has been similar. To achieve its objectives, NNP implemented various interventions targeted at children under two years of age, adolescent girls (aged 13–19 years), newly married couples, and pregnant and lactating women. NNP included a core package of area-based community nutrition services:

- Behavior change and communication (BCC) at the community and household level to address maternal, infant, child, and adolescent feeding, and caring practices impacting nutrition
- Growth monitoring and promotion
- Micronutrient supplementation (vitamin A for children aged 9–59 months and iron-folate for pregnant women)
- Biannual deworming of severely malnourished children (aged 12–59 months) and adolescents (aged 13–19 years)
- Utilization of nutrition, health, and food security services
- Food supplementation (*pushti* packets) for severely malnourished children under two years of age
- Gardening and poultry activities to improve food security (the gardening and poultry activities were discontinued in 2006).

The method of service delivery under BINP, NNP, and HNPSP was area-based community nutrition activities contracted out to nongovernmental organizations (NGOs). Community nutrition activities were organized around community-donated Nutrition Centres, established for a population of 1,250 to 1,500, and run by part-time female workers, called Community Nutrition Promoters (CNPs). The CNPs were supervised by Community Nutrition Organizers. With financing from HNPSP, the Institute of Public Health Nutrition (IPHN), under the Directorate General of Health Services of MOHFW, provided micronutrient supplementation throughout the country. Other national-level nutrition activities consisted of communication support (implemented by United Nations Children's Fund [UNICEF]), breastfeeding promotion and support (implemented by the Bangladesh Breastfeeding Foundation), and iodine fortification of salt.

Evaluation of the nutrition activities under HNPSP revealed these were successful in improving knowledge and feeding practices of the households (Mbuya and Ahsan 2013) but did not have any significant impact in reducing undernutrition. With the closure of HNPSP in June 2011, the MOHFW started implementing a follow-on program, the Health, Population and Nutrition Sector Development Programme (HPNSDP, 2011–16). Building on the lessons learnt under HNPSP, the program is costed at US$7.7 billion and is cofinanced by the GOB and DPs. In an effort to accelerate progress in reducing the persistently high rates of maternal and child undernutrition, in June 2011 the GOB committed to mainstream and scale up HPNSDP's essential nutrition interventions into the existing health and family planning services.

Under HPNSDP, IPHN has been mandated to lead the nutrition component, called the National Nutrition Services (NNS), with a total budget allocation of approximately US$190 million for the period July 2011 to June 2016. The main services being provided through the various tiers of health and family planning facilities include treatment and referral of severe cases of undernutrition, BCC, screening of undernutrition, promotion of infant and young feeding practices, and micronutrient supplementation.

In 2014, the World Bank and NNS commissioned an operations research to assess (i) the effectiveness of the delivery of the different components of NNS, and (ii) whether the various interventions are being delivered to the intended beneficiaries with adequate coverage and quality (World Bank 2014). The assessment identified several substantial challenges in the management and implementation of NNS related to delivery and intervention platform choices, governance and institutional choices, training and rollout, and service delivery. The assessment concluded that the overall NNS effort is an ambitious, but valuable, approach to examining how best to support nutrition actions through an existing health system with diverse platforms. To achieve the desired level of impact and coverage, it is necessary to address critical challenges related to leadership and coordination and implement a prioritized set of interventions with well-matched (for scale, target populations, and potential for impact) health system delivery platforms. Strategic investments in ensuring transparency, and engaging available technical partners for monitoring and implementation support, could also prove fruitful. Although the health sector in particular must lead the effort on nutrition, other key sectors could coordinate their own "nutrition-sensitive" policies and programs and provide the necessary support.

Hygiene practices are embedded in the various interventions of HPNSDP as BCC activities as mentioned in the Strategic Document of HPNSDP. The National Health Policy 2011, however, does not highlight the importance of hygiene in improving HNP outcomes. All HNP policies and strategies need to single out hygiene as a cross-cutting area and prioritize it as one of the main pillars for intervention. Without this, promotion of hygiene practices will not get due attention and will remain a side issue. The health and family planning workers also will not be able to fully appreciate the critical role of hygiene.

The Case for a Coordinated Multisectoral Response to Undernutrition in Bangladesh

Undernutrition is a multifactorial challenge—a consequence of factors operating at several levels and across multiple sectors. The potential causes of undernutrition may be classified as *immediate, underlying,* or *basic.* Figure 2.5 presents a conceptual framework depicting the causes of child undernutrition, which is

Figure 2.5 Conceptual Framework for the Causes of Undernutrition

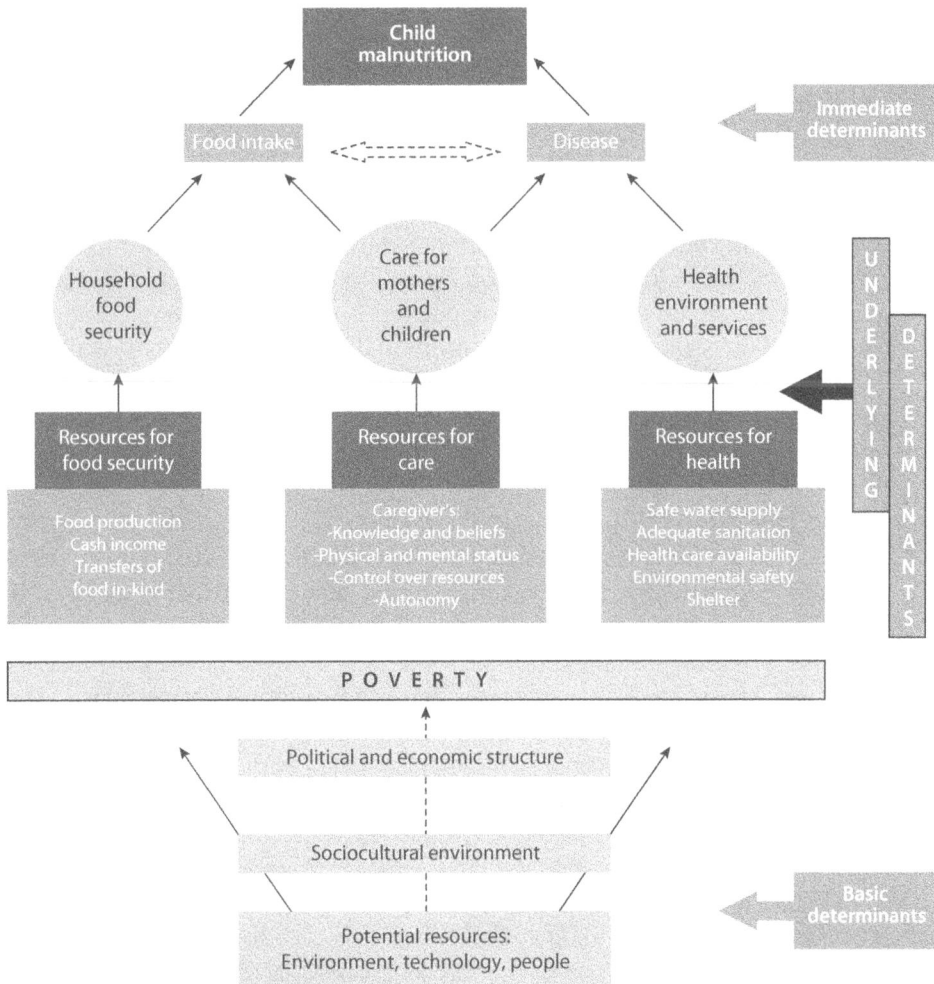

Sources: Adapted from UNICEF; Engle, Menon, and Haddad 1999; Smith and Haddad 2000.

adapted from UNICEF and subsequent work in this area (Engle, Menon and Haddad 1999, and Haddat et al. 2002). This framework highlights the need to work in multiple sectors in order to address the problem of undernutrition.

Inadequate dietary intake and disease are often the *immediate* causes of undernutrition and directly affect the individual. Moreover, they form a vicious cycle: Inadequate dietary intake increases the likelihood of illness because of weakened immune levels; illnesses lead to a loss of appetite and poor absorption, which in turn worsen undernutrition.

The main *underlying* causes of undernutrition are lack of household food security, inadequate care for mothers and children, and poor health and environmental conditions. Each factor is determined by the social and economic resources available to the individuals and the household as a whole. Poverty is a key factor affecting all underlying determinants.

Caring practices include appropriate nutrition and support for mothers during pregnancy and lactation, infant feeding practices (breastfeeding and complementary feeding), and health-seeking behaviors and cognitive stimulation. The caregiver's knowledge and beliefs also are important resources that influence what types of health services are accessed and what caring practices are adopted.

Factors affecting the health and environment conditions of the household include access to health care from affordable, qualified providers and safe water and sanitation services. Poor environmental safety, including lack of adequate shelter, is also a critical determinant of undernutrition.

The *basic* causes of undernutrition are insufficient resources available at the country or community level, and the political, social, and economic conditions that govern how these resources are distributed. The basic causes also influence institutions. These include both the formal institutions that provide public sector services, such as health and education, and the informal institutions that determine the social and cultural norms regarding the rights of women and vulnerable groups in the population.

The causes of undernutrition in Bangladesh are multifactorial as discussed above. A Helen Keller International study (HKI 2006) reported that the most important explanatory variables for stunting among under-five children in Bangladesh included food intake, household food insecurity, poor maternal and childcare practices, disease, and limited access to a healthy environment (safe water and sanitation). The study also found that stunting was very high even among the wealthiest groups, an indication that economic growth alone is not sufficient to improve nutrition, a point highlighted earlier in this report. The multidimensional nature of the causes of undernutrition in Bangladesh underscores the diversity of actions that are needed across sectors, levels, and actors to address the problem. Although the HNP sector in Bangladesh continues to and should play a central role in delivering direct nutrition interventions through the maternal, neonatal, and child health services, the GOB also recognizes that health sector–based nutrition programs, though essential, have not been and will not be adequate in reducing Bangladesh's very high levels of maternal and child undernutrition.

There is now substantial global evidence showing that direct actions to address the immediate determinants of undernutrition (nutrition specific) can be enhanced by actions addressing the more underlying determinants (nutrition sensitive), which are in the domains of ministries other than health, hence the need for a multisectoral approach (Gillespie et al. 2013). Such actions can strengthen nutritional outcomes in three main ways by: (i) accelerating action on determinants of undernutrition; (ii) integrating nutrition considerations into programs in other sectors that may be substantially larger in scale; and (iii) increasing "policy coherence" through governmentwide attention to nutrition.

While the need for a multisectoral response to undernutrition has long been recognized, the required institutional arrangements are not clear. In the 1970s, multisectoral nutrition planning cells were introduced in many countries and placed centrally in a planning commission, or in the Office of the President (Levinson and McLachlan 1999). The planning cells were expected to be able to affect a broad range of development policies and programs as a result of their high level placement. The U.S. Agency for International Development (USAID) and the Food and Agriculture Organization (FAO) supported the establishment of 26 nutrition planning cells in developing countries throughout the 1970s (Levinson 1999; and Rokx 2000). The BNNC described above was set up as part of this global push for multisectoral nutrition planning cells to coordinate nutrition policy.

The design and implementation of multisectoral strategies to address undernutrition have been far from successful in Bangladesh and around the world. The nutrition planning cells initiated in the 1970s had no significant impact. They lack the authority and resources to coordinate the relevant sectors effectively or to introduce incentives to promote cross-sectoral coordination. The BNNC was no different.

A more realistic and workable institutional arrangement is to equip the different sectors with the required latitude and resources to carry out their own programs. The nutrition coordination agency can be granted the authority to define overall policies and guide the allocation of resources (Heaver 2005). The coordination agency's role would be to ensure that correct incentives are in place to motivate sector agencies to prioritize nutrition, to operate accountability mechanisms to ensure that the sectoral agencies do carry out their nutrition functions, and to engage in sectoral policy design and implementation to ensure that undernutrition remains a priority (World Bank 2006).

The World Bank has commissioned a series of analytical works to identify the contribution of sectors (other than health) to undernutrition. The first two reports in this series considered the impact of agriculture and microcredit, and gender, on nutrition. The first report documented that in the agriculture sector, the greatest potential for improving outcomes by integrating nutrition lies in the production of fruit, vegetables, livestock, and aquaculture, particularly small-scale production at the household level. In microcredit, the greatest potential lies in targeting microcredit programs that women are most likely to be involved in, and where there is a strong income effect.

The second report examines the role of gender in affecting undernutrition outcomes and how gender constructs determine undernutrition outcomes in Bangladesh. The report was based on an analysis of two data sets with detailed information on nutrition and gender relations within the household: the nationally representative 2007 BDHS (NIPORT et al. 2007), and a longitudinal data set, spanning 10 years (1996/97–2006/07), that was collected as part of an International Food Policy Research Institute (IFPRI) study on micronutrient and gender impacts of agricultural tecnhnologies. The study found that the impact of women's mobility, decision making, and other measures of women's empowerment on child nutritional status were weak or inconsistent. However, experience of any form of domestic partner abuse was strongly associated with adverse nutrition outcomes for women and their children. As such, the report concluded that primary prevention approaches for intimate-partner violence and sexual violence should be prioritized by relevant government departments, together with strengthening effective legal protection against all forms of domestic violence and sexual abuse.

This report, the third in the series, examines the role of the water and sanitation sector in improving nutrition outcomes. The objective is to identify potential sector-specific resources that could be mobilized to alleviate undernutrition in Bangladesh.

CHAPTER 3

How Water and Sanitation Can Improve Nutrition Outcomes

Key Messages

- Evidence of the impact of poor water and sanitation on diarrhea is undisputable, and over the last few decades, diarrhea has also been implicated as an important cause of poor infant and child growth. However, recent evidence suggests that the effect of diarrhea on permanent stunting is smaller than previously thought.
- Through an extensive oral rehydration program, Bangladesh has been successful in managing diarrhea, which has contributed to a sharp decline in child mortality. However, this has not translated to a comparable effect on nutrition outcomes. This observation also helps in interpreting the weak linkage of diarrhea to undernutrition.
- A hypothesis by Humphrey (2009) posits that the predominant impact of contaminated water and poor sanitation on undernutrition is via tropical/environmental enteropathy rather than diarrhea.
- Both the diarrheal and the tropical/environmental enteropathy hypotheses are premised upon fecal-oral contamination. However, it is the biological "response" to the fecal-oral contamination that is different.
- Diarrhea is a clinical condition and results in loss of appetite and nutrients, while tropical/environmental enteropathy is subclinical (without signs/symptoms). It is characterized by physiological and anatomical changes to the structure of the small intestine that affect a child's ability to absorb and utilize nutrients.
- To have a sustained impact on undernutrition, improved water, sanitation, and hygiene (WASH) interventions are necessary, but not sufficient. Bangladesh needs, at a minimum, to ensure adequacy in three dimensions—availability of food, health care, and environmental health for all—in order to tackle the problem of undernutrition (Newman 2013). Sufficiency in just one of these sectors will have a sustained impact on undernutrition (Newman 2013).

What Are the Pathways of Influence between Water and Sanitation and Nutrition?

Diarrhea as Both a Cause and Effect of Undernutrition

Diarrhea and undernutrition, alone or together, constitute major causes of morbidity and mortality among children throughout the world. Scrimshaw, Taylor, and Gordon (1968) presented the synergistic relationship of undernutrition and infectious diseases. Infections have more serious consequences in malnourished people, and, conversely, infectious diseases can result in borderline nutritional deficiencies becoming more severe undernutrition. An early estimate of the World Health Organization (WHO) showed that almost half of undernutrition in the world was associated with repeated diarrhea or intestinal worm infections. These were caused by unsafe water, inadequate sanitation, or insufficient hygiene. Subsequent studies from various countries suggested that diarrheal illnesses affect a child's growth by reducing gains in weight and height of a child (Guerrant et al. 1992). They concluded that the greatest effects of diarrhea are witnessed with frequent/recurrent bouts of the illness, which reduce the critical catch-up growth that otherwise occurs after diarrheal illnesses or severe undernutrition. Analyses from Northeast Brazil (Guerrant et al. 1992) indicated that undernutrition can lead to a 37 percent increase in frequency and a 73 percent increase in duration of diarrheal illnesses, accounting for a doubling of the diarrhea burden (days of diarrhea) in malnourished children.

The concept of diarrhea causing and being a consequence of undernutrition has also evolved over time. Brown (2003) compiled various studies on diarrhea and nutrition undertaken from 1968 to 1998 documenting the impact and risk factors of diarrhea. Brown found an intertwined relationship between diarrhea and undernutrition: Children with diarrhea eat less and are not fully able to absorb the nutrients from their food; while malnourished children are more vulnerable to diarrhea (compared to normal children) when exposed to fecal material from their environment. Brown concluded that infection adversely affects nutritional status by reducing intake of food, lowering absorption capacity of the intestine, increasing catabolism,[1] and taking away nutrients from the body that are required for growth. Furthermore, undernutrition reduces the protection of the body against infection and alters the immune function, thereby prompting infection.

Martorell, Yarbrough, and Klein (1980), Rowland, Coal, and Whitehead (1977), and Black, Brown, and Becker (1984) developed statistical models based on data from Guatemala, West Africa, and Bangladesh, respectively, to estimate the proportion of the total growth deficit that could be attributed to diarrhea. They concluded that as much as one-fourth to one-third of the observed growth failure could be attributable to enteric infections. Martorell, Yarbrough, and Klein (1980) noted that fully weaned Guatemalan children reduced their energy intake by almost one-third during acute infections. However, Brown et al. (1985) suggested that the reduction of energy intake caused by diarrhea was partially prevented by breastfeeding based on data collected from Bangladesh. They found that Bangladeshi children who were still breastfeeding reduced their intakes by only about 7 percent; while intake of nonbreast-milk energy declined by about 30 percent during illness, there were no changes in breast milk consumption. Rowland, Coal, and Whitehead (1977) also found that the previously observed diarrhea-induced growth deficit was absent in fully breast-fed infants in an urban field site in West Africa, and they concluded that exclusive breastfeeding extends protection from the adverse nutritional consequences of diarrhea.

Subsequent studies by Brown et al. (1989) in Peru, Popkin et al. (1990), and Kramer et al. (2001) in Belarus found that exclusively breast-fed infants, compared with infants who either received other foods or liquids along with breast milk or were fully weaned from the breast, had considerably reduced risks of diarrhea (and other infections). There is also some evidence suggesting that vitamin A reduces the severity of diarrheal illness but has no effect on the incidence. There is evidence that zinc supplementation can reduce the incidence of diarrhea by almost 20 percent (Brown 2003).

The weak linkage between diarrhea and undernutrition assists in interpreting the successful management of diarrhea in Bangladesh. Through an extensive oral rehydration program and very high coverage of vitamin A supplementation across the country (figures 3.1 and 3.2), Bangladesh has been successful in managing diarrhea. As shown by the Bangladesh Demographic and Health Survey (BDHS), the prevalence of diarrhea among children below the age of five years declined significantly from 12.6 percent in 1993–94 to 4.6 percent in 2011 (NIPORT et al. 1993–94 and 2011). This is mirrored by a marked decline in child mortality from 133 deaths per 1,000 births in 1993–94 to 53 deaths in

Figure 3.1 Prevalence and Treatment of Diarrhea in Bangladeshi Children Aged Less Than 5 years, 1993–2011 (%)

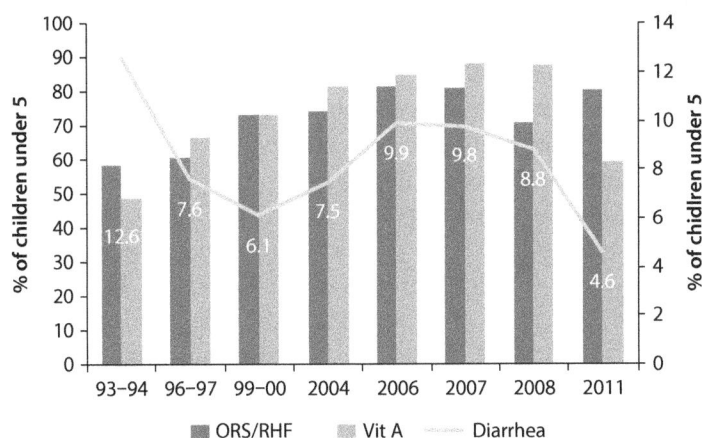

Source: NIPORT et al. 2013, various years.
Note: Data for 1993–94 is for children aged less than three years.

Figure 3.2 Trends of Mortality and Prevalence of Diarrhea in Bangladeshi Children Aged Less Than 5 years, 1993–2011 (%)

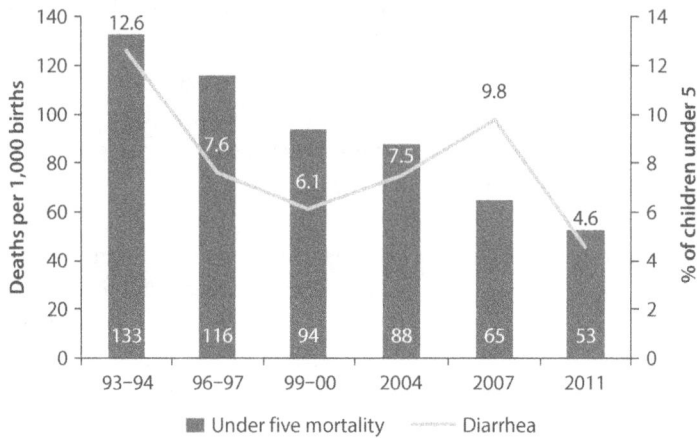

Source: NIPORT et al. 2013, various years.
Note: Data for prevalence of diarrhea for 1993–94 is for children aged less than three years. Under-five mortality is the probability of dying between birth and the fifth birthday.

2011 (NIPORT et al. 1993–94 and 2011). However, these trends are accompanied by little effect on nutrition outcomes. As discussed in chapter 2, in 2013 the rates of undernutrition in Bangladesh (underweight rate of 35 percent and stunting 39 percent) remain among the highest in the world (NIPORT 2013).

The Emergence of the Tropical/Environmental Enteropathy Hypothesis

Humphrey (2009) hypothesized that "*the primary causal pathway from poor sanitation and hygiene to under-nutrition is tropical enteropathy and not diarrhoea.*" She noted that "*a key cause of child under-nutrition is a subclinical disorder of the small intestine known as tropical enteropathy, which is characterised by villous atrophy, crypt hyperplasia, increased permeability, inflammatory cell infiltrate, and modest malabsorption.*" Basically this means that because of chronic exposure to (mostly) fecal bacteria, the structure (decrease in the villous height) and function of the small intestine changes, which initiates a sequel leading to undernutrition. Villi are small fingerlike projections present in the lining of the small intestine. Digestion largely occurs in the ileum of the small intestine and from there the digested end products (glucose, amino acids, and so forth) move into the blood through a process called absorption. In order to make absorption quicker and more efficient, the ileum walls need to have a large surface area, which is provided by villi. Decreased villous height reduces the total area of the small intestine and the absorption of nutrients. In addition, increased permeability of the intestinal tract affects the ability of the body to prevent pathogens from breaching the intestinal barrier. This triggers the body's immune response, resulting in nutrients being prioritized for defense rather than normal growth.

Humphrey further stated that "*tropical enteropathy is caused by faecal bacteria ingested in large quantities by young children living in conditions of poor sanitation and hygiene.*" Such conditions are linked to poor sanitation and hygiene practices which are prevalent in developing countries. Humphrey suggested that greater "*provision of toilets and promotion of handwashing after faecal contact could reduce or prevent tropical enteropathy and its adverse effects on growth.*"

Figure 3.3 illustrates the pathways proposed by Humphrey (2009). Children living in household with poor sanitation facilities are exposed to high concentrations of fecal bacteria and end up ingesting more bacteria than children living in relatively cleaner households. The fecal bacteria colonize the small intestine and induce tropical/environmental enteropathy. These changes, coupled with reduced nutrient absorption, marginal dietary intake, and the high growth demands of the first two years of life, cause growth faltering.

Lin et al. (2013) undertook a study in rural Bangladesh that lends support to Humphrey's hypothesis. They suggested that "*children living in clean households with good hygiene would have lower prevalence of parasites and environmental enteropathy and better growth (less stunting, wasting, and underweight conditions) compared with children living in contaminated households with poor hygiene.*" Consistent with this supposition, they found that children in environmentally clean households had lower levels of parasitic infection, gut function, and

Figure 3.3 Pathways for Tropical Enteropathy

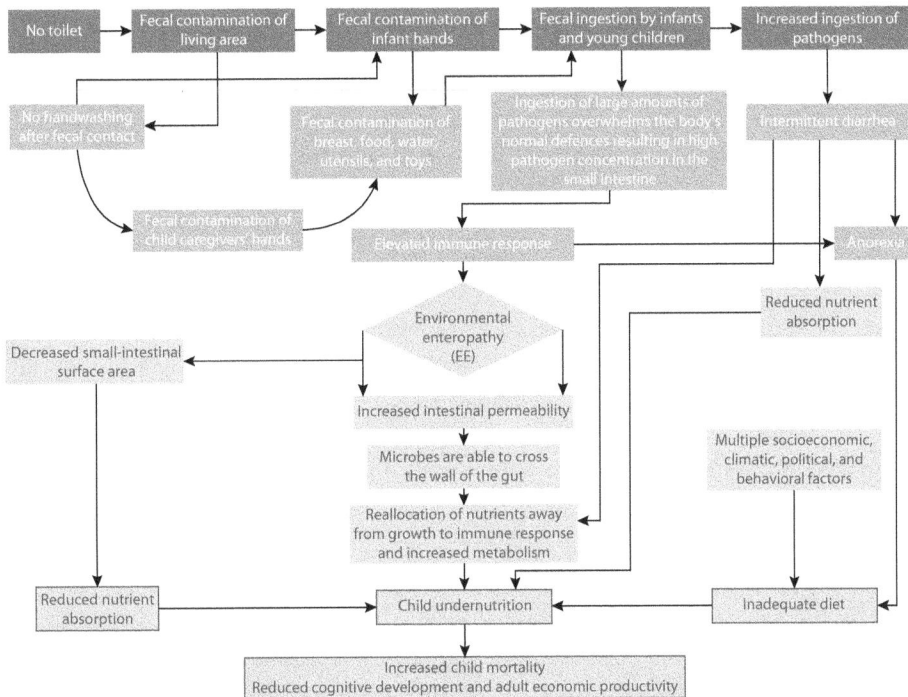

Source: Humphrey 2009.

growth compared with children in contaminated households, even in the absence of drastic infrastructure improvements. The prevalence of stunting was 22 percent lower among children living in clean households compared with children living in contaminated households.

An earlier study conducted by Campbell, Elia, and Lunn (2003) in The Gambia had similar findings, estimating that environmental enteropathy explained 40–64 percent of stunted growth in a small cohort of children. The study by Lin et al. (2013) had some limitations: It had a small sample size, was observational, and the results, while plausible, were perhaps subject to bias. Additionally, the study lacked the temporal ordering necessary to establish causal relationships and the improvements documented by Lin et al. were possibly brought about by prenatal or early postnatal interventions, such as maternal and child nutrition, which they did not measure in the study. These effects cannot be excluded from the study cohort, as differences in stunting between clean and contaminated environments were already in place by 2007, but environmental enteropathy markers were not measured until 2010.

What Is the Impact of Water and Sanitation Interventions on Nutrition Outcomes?

Water, sanitation, and hygiene (WASH) interventions have traditionally been implemented to reduce infectious diseases, and sometimes with a view of subsequently improving nutritional conditions. There is, however, limited evidence of WASH interventions alone improving nutrition.

In a Cochrane Review, Dangour et al. (2013) completed a meta-analysis of 14 studies from 10 low- and middle-income countries to evaluate the effect of WASH interventions on the nutritional status of children. The review included randomized and nonrandomized interventions, or a combination of interventions, for children aged less than 18 years. The interventions were designed to (i) improve the microbiological quality of drinking water or protect the microbiological quality of water prior to consumption; (ii) introduce new or improved water supply or improve distribution; (iii) introduce or expand the coverage and use of facilities designed to improve sanitation; and (iv) promote handwashing with soap after defecation and disposal of child feces, and prior to preparing and handling food. The authors concluded that WASH interventions may slightly improve the height of children under five years of age. The conclusions are based on relatively short-term studies and, therefore, need to be treated with caution.

Dangour et al. (2013) also state that,

"there is suggestive evidence from cluster-randomized controlled trials of a small benefit of WASH interventions on measures of growth in childhood. There is no evidence of the effect of other WASH interventions on nutritional outcomes in children and a major gap in the literature concerns the effect of water supply and sanitation interventions on nutrition outcomes. Non-randomized studies provided mixed evidence on the effect of a

variety of WASH interventions on nutrition outcomes. All interventions were conducted in children under the age of five years and there is no evidence of the effect of WASH interventions in children older than five years of age."

Clasen et al. (2014) report that it cannot be inferred that increasing the coverage of latrines can effectively reduce exposure to fecal pathogens if people do not actually use the latrines. They conducted a cluster-randomized trial in 100 rural villages of Odisha, India, to test the effectiveness of rural sanitation program on diarrhea, soil-transmitted helminth infection, and child undernutrition. Clasen et al. found that programs focusing on latrine construction do not actually change behavior and thus do not reduce exposure to pathogens. The authors suggest that to improve sanitation conditions, the interventions should be implemented in a way that ensures uptake of the WASH facilities, reduces exposure to fecal matter, and demonstrates health gains. A target of only increasing coverage of WASH interventions may not be effective.

Newman (2013) concluded that adequate food, environmental health, and care is associated with considerably lower levels of stunting than adequacy in none or only one of the dimensions (after controlling for wealth, education of the mother, and other control variables). This supports the notion that a multisectoral approach is needed to address stunting. Newman (2013) showed strong correlations between stunting and food, environmental health, and care adequacy for Bangladesh, Peru, and India, countries with differing average levels of stunting. For example, the author shows differences in stunting between Indian children who have adequate food, environmental health, and care and children who are inadequate in all three dimensions: 30 percentage points without controls and 14–22 percentage points with controls. These associations could be further investigated through randomized control trials.

The findings relating to Bangladesh are based on analysis of a data set from the Bangladesh Food Security and Nutritional Surveillance Survey (FSNSP) conducted by Helen Keller International, the James P. Grant School of Public Health of BRAC University, and the Bangladesh Bureau of Statistics. Adequate food included parameters of mother's dietary diversity, exclusive breastfeeding of children, food never being restricted to young child, and household food insecurity access score. Adequate environmental health included improved water source, improved sanitation source, and handwashing. Care practices included measure of antenatal care visits, immunization of children, duration of breastfeeding of children, mother receiving iron-folic acid supplementation, mother's body mass index, and the proximity to a health clinic.

Figure 3.4 shows that in 2013, Bangladeshi children who were adequate in all three dimensions—food, environmental health, and care—were less likely to be stunted. The stunting rates of children who are adequate in all three dimensions is almost 30 percentage points lower than the rates of those children who are inadequate in all dimensions. There is also a notable difference in stunting rates across children who are adequate in three, two, or one dimensions.

Figure 3.4 Percentage of Bangladeshi Children (Aged 6–60 months) Who Are Stunted by Adequacy of Food, Environmental Health, and Care, 2013

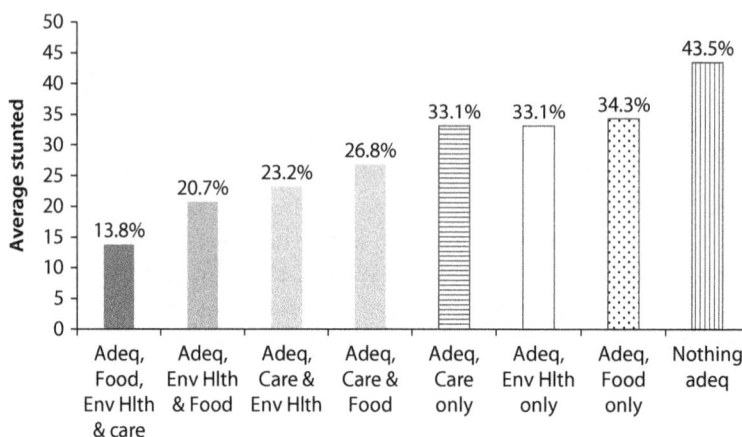

Source: Newman 2013.
Note: Adeq. = adequate; Env Hlth = environmental health.

Figure 3.5 Percentage of Bangladeshi Children within Each Category of Adequacy of Food, Environmental Health, and Care, 2013

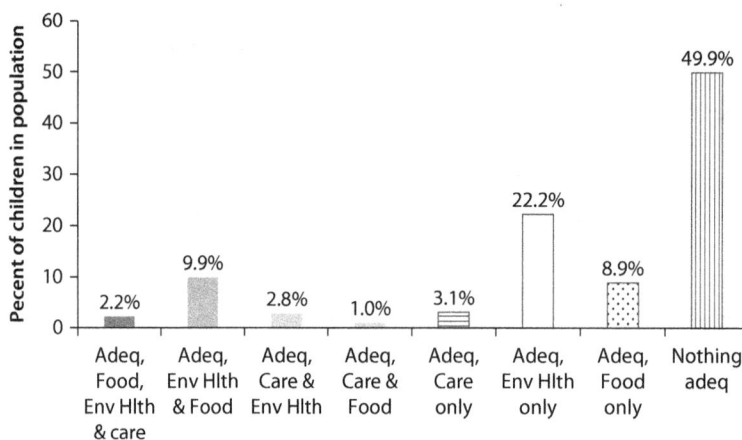

Source: Newman 2013.
Note: Adeq. = adequate; Env Hlth = environmental health.

Figure 3.5 shows that very few children (2.2 percent) in Bangladesh actually are adequate in all three dimensions. Almost 50 percent of the children are inadequate in all dimensions.

Figures 3.4 and 3.5 show the results for all children in the sample without taking into consideration their family's wealth. Figures 3.6 and 3.7 present the stunting rates for the same eight categories of adequacy of food, environmental health, and care by wealth terciles. Figures 3.6 and 3.7 show that inadequacy on

Figure 3.6 Percentage of Bangladeshi Children (of Poorest and Middle Wealth Terciles) Who Are Stunted by Adequacy of Food, Environmental Health, and Care, 2013

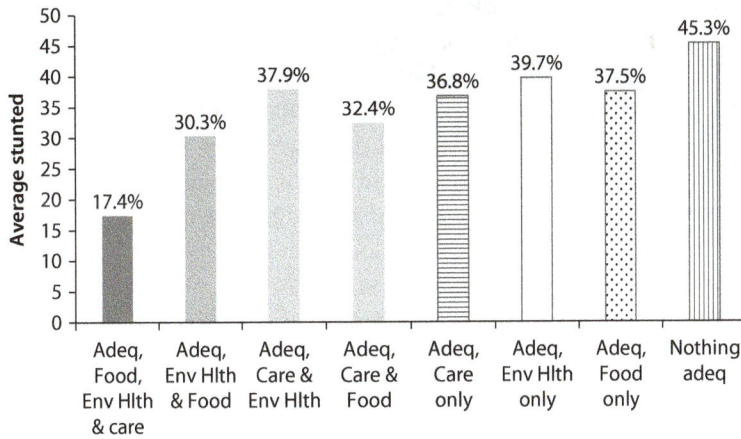

Source: Newman 2013.
Note: Adeq. = adequate; Env Hlth = environmental health.

Figure 3.7 Percentage of Children (of the Wealthiest Tercile) Who Are Stunted by Adequacy of Food, Environmental Health, and Care, 2013

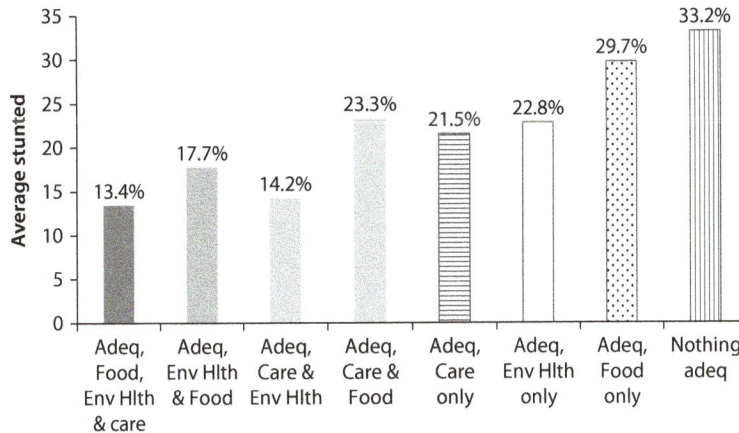

Source: Newman 2013.
Note: Adeq. = adequate; Env Hlth = environmental health.

all dimensions is considerably higher (roughly 40 percentage points higher) for the lowest two wealth terciles than for the wealthiest tercile. Moreover, the same patterns of stunting hold when data are filtered by wealth category: There are considerably lower stunting rates when the children are adequate in all dimensions than when the children are adequate in none.

The overall conclusion that can be drawn from these findings, and those of other studies described in this report, is that a multisecoral approach that builds on established conceptual frameworks is needed to achieve meaningful gains in nutrition outcomes. Availability and consumption of nutritionally adequate (in terms of quality and quantity) food is essential for a child's growth. Clean water and appropriate sanitation are also necessary to prevent children from experiencing frequent illnesses and losing nutrients and subsequently faltering in their growth. Therefore, in order to make a significant dent in the high levels of undernutrition, Bangladesh needs at a minimum to ensure adequacy in all three dimensions of food, environmental health, and care for all. Building a toilet will not by itself translate into the growth of a child; food alone might not be adequately absorbed and utilized. These various dimensions are necessary, but alone are not necessarily sufficient.

In line with the focus of this report, Chapter 4 documents the efforts of the GOB in the water and sanitation sector.

Note

1. Catabolism is a set of metabolic activities that break down larger molecules into smaller units.

Achievements in the Water and Sanitation Sector

Key Messages

- Bangladesh has made significant progress in increasing coverage of water and sanitation facilities. In 2012, open defecation has been reduced to 3 percent of the population while improved water supply is being provided to 85 percent of the population.
- Of the total funds available for water and sanitation, 35 percent is from Government of Bangladesh (GOB) sources, with the majority (65 percent) from development partners (DPs) and nongovernmental organizations (NGOs). To sustain the progress achieved so far, a greater contribution from GOB sources is required.
- A key achievement in sanitation has been the shift from open defecation to "fixed point defecation." But quality of the sanitation facilities remains an issue, with only half the population using improved sanitation facilities.
- In water, progress has been made in transitioning from traditional drinking sources to piped/improved sources. The coverage of piped water supply needs to be further expanded (as currently only one-third of the population has access to piped water) and the quality of water improved, among other things.
- Almost all of the interventions in the water and sanitation sector have focused on building infrastructure for providing access to water and sanitation facilities. The "softer" sides of the sector (quality, hygiene, innovation, and so forth) have not progressed as much. Lack of hygiene practices is a major concern, and the GOB has developed strategies for addressing this.
- The significant decline in the rates of open defecation cannot be considered as a "mission accomplished." Unless the transmission of infections can be limited, further investments in increasing the coverage of water supply and sanitation facilities will not yield adequate results. An emphasis on improving hygiene practices has to be prioritized, along with innovation for designing "improved" and affordable sanitation facilities.

- Bangladesh has formulated a set of comprehensive policies and strategies, covering almost all of the issues facing the sector—four legislative acts, two national policies, and five national strategies (excluding the draft 2014 Strategy). However, translating these policies into action has not progressed much. Moreover, these policies and strategies have focused on expanding coverage of water and sanitation facilities without a corresponding focus on hygiene.
- In 2014, the GOB drafted the National Water Supply and Sanitation Strategy, which provides an all-encompassing policy direction for the sector. The draft 2014 Strategy focuses on improving the quality and coverage of water and sanitation facilities, while prioritizing hygiene practices and the critical role of the health and education sectors in ensuring effective implementation of water, sanitation, and hygiene (WASH) interventions. It is important that the 2014 Draft Strategy be finalized and the time-bound action plan be implemented. These actions will contribute to the national effort of for improving WASH conditions and, hence, reducing undernutrition.

The Status of Water and Sanitation in Bangladesh

Bangladesh is on track to achieve the Millennium Development Goal (MDG) target for water, but is not on track for the target on sanitation defined in MDG 7c: "*Halve, by 2015, the proportion of people without sustainable access to safe drinking-water and basic sanitation.*" This subsection documents the progress achieved in the provision of water and sanitation facilities and highlights the key challenges.

Progress in Sanitation—From Open Defecation to "Fixed Point Defecation"

Improvements have been made in coverage of sanitation facilities for the population. In 2012, 97 percent of the population had access to a latrine facility (irrespective of their quality, and including pit latrines, household latrines with septic tanks, community sanitation facilities, and small-bore sewerage systems); only 3 percent practiced open defecation (as shown in table 4.1). Improved sanitation facilities are available to 57 percent of the population (WHO-UNICEF Joint Monitoring Programme, JMP 2014).

Table 4.1 Improvements in Sanitation Facilities, 1990–2012

	Urban (%)		Rural (%)		Total (%)		
Sanitation coverage	1990	2012	1990	2012	1990	2012	2015
Improved facilities	46	55	30	58	33	57	70
Shared facilities	25	30	15	28	17	28	—
Other unimproved	19	15	15	11	16	12	—
Open defecation	10	0	40	3	34	3	—

Sources: Figures for 1990 and 2012 taken from JMP 2014; MDG target for 2015 taken from Water and Sanitation Sector Local Consultative Group 'Open Data' Monitoring Platform.
Note: — = Not available.

An improved sanitation facility is defined by JMP as one that hygienically separates human excreta from human contact. To allow for international comparability of estimates, JMP uses the classification shown in table 4.2 to differentiate between "improved" and "unimproved" types of sanitation facilities.

According to a national baseline survey conducted by the GOB in 2003, one-third of the population (33 percent) was using hygienic latrines, around 25 percent were using unhygienic hanging latrines, and almost half the population (42 percent) did not have access to any kind of latrine facility and were resorting to open defecation. This led the GOB to launch the National Sanitation Campaign, which resulted in rapid progress (more than 9 percent progress per year) in sanitation coverage. Consequently, open defecation has come down remarkably from 42 percent of the population in 2003 to 3 percent in 2013 (JMP 2014). More importantly, Bangladesh has successfully reduced the percentage of the bottom 40 percent of the population without access to basic sanitation facilities (figure 4.1).

Table 4.2 JMP Definition of Improved and Unimproved Sanitation

Improved sanitation	Unimproved sanitation
• Use of	• Use of
o Flush or pour-flush to	o Flush or pour-flush to elsewhere (excreta are flushed
o piped sewer system	to the street, yard or plot, open sewer, a ditch, a
o septic tank	drainage way, or other location)
o pit latrine	o Pit latrine without slab or open pit
• Ventilated improved pit latrine	o Bucket
• Pit latrine with slab	o Hanging toilet or hanging latrine
• Composting toilet	• Public or shared sanitation facilities
	• No facilities or bush or field (open defecation)

Source: JMP 2014.

Figure 4.1 Access to Sanitation Facilities in Bangladesh by Income Quintile, 1995 and 2008 (%)

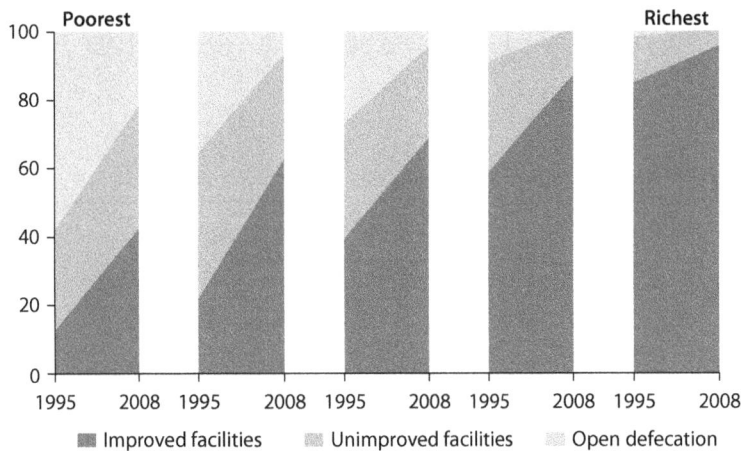

Source: JMP 2013.

The success achieved so far in sanitation (particularly the significant reduction of open defecation) in Bangladesh is largely credited to the Community-Led Total Sanitation (CLTS), which was developed in Bangladesh by WaterAid and the concept adopted by the GOB in the early 2000s. The approach is led by and works for whole communities rather than individuals. Under the CLTS approach, communities are facilitated to conduct their own appraisal and analysis of open defecation and take necessary actions to achieve "open defecation free" status. CLTS emphasizes the behavioral change that is needed to ensure real and sustainable improvements. CLTS works to raise awareness that as long as even a minority of the community continues to defecate in the open everyone is at risk of disease. In this way, the approach triggers the community's desire for collective change, motivates people into action, and encourages innovation, mutual support, and appropriate local solutions, thus leading to greater ownership and sustainability.

Progress in Provision of Water—From Traditional Sources to Piped/improved Sources

With regards to water, 85 percent of the Bangladeshi population has access to improved water supply—with the vast majority through nonpiped, "other improved" sources (JMP 2013) (table 4.3). Piped water supply is mainly found in the urban areas. In the rural areas, water supply is primarily through hand-pump tubewells and other water points like pond sand filters, ring wells, and rainwater harvesting units (LGD 2011b).

The JMP defines an improved drinking water source as one that, by nature of its construction or through active intervention, is protected from outside contamination, in particular from contamination with fecal matter. To allow for international comparability of estimates, JMP uses the classification shown in (table 4.4) to differentiate between "improved" and "unimproved" drinking water sources.

The other achievement with regards to water is that Bangladesh has successfully increased access of the bottom 40 percent of the population to basic water services, similar to the increase in the provision of sanitation facilities (JMP 2013) (figure 4.2).

Table 4.3 Improvements in Coverage of Water Source in Bangladesh, 1990–2012 (%)

	Urban (%)		Rural (%)		Total (%)		
Drinking water source	1990	2012	1990	2012	1990	2012	2015
Piped onto premises	23	32	0	1	5	10	89
Other improved	58	54	65	83	63	75	
Other unimproved	17	14	28	16	26	15	–
Surface water	2	0	7	0	6	0	–

Source: Figures for 1990 and 2012 taken from JMP 2014; MDG target for 2015 taken from Water and Sanitation Sector Local Consultative Group "Open Data" Monitoring (Web-based) Platform.

Table 4.4 JMP Definition of Improved and Unimproved Drinking Water

Improved drinking water	*Unimproved drinking water*
Use of	Use of
• Piped water into dwelling, plot, or yard	• Unprotected dug well
• Piped water into neighbor's plot	• Unprotected spring
• Public tap/standpipe	• Small cart with tank/drum
• Tubewell/borehole	• Tanker truck
• Protected dug well	• Surface water (river, dam, lake, pond, stream, channel, irrigation channel)
• Protected spring	• Bottled water (considered to be improved only when the household uses water from another improved source for cooking and personal hygiene; where this information is not available, bottled water is classified on a case-by-case basis)
• Rainwater	

Sources: JMP 2014.

Figure 4.2 Access to Water in Bangladesh by Income Quintile, 1995 and 2008 (%)

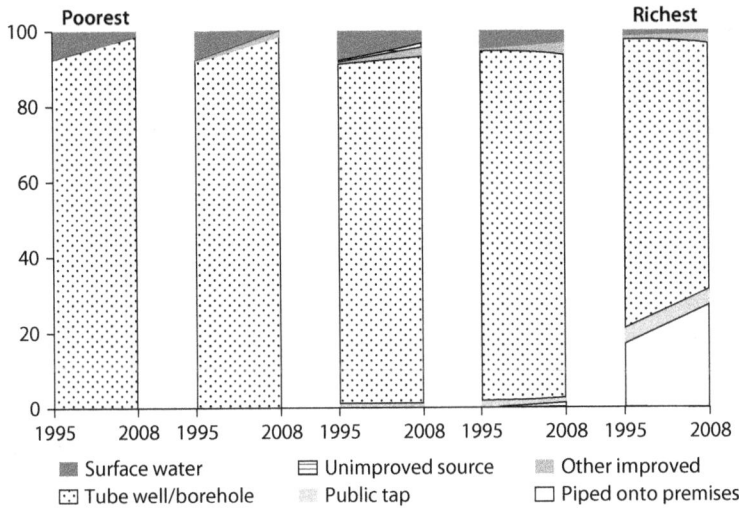

Source: JMP 2013.

The Sector Development Plan 2011–25 (LGD 2012) shows that prior to 1971, the traditional sources of drinking water were mainly ponds, dug wells, and canals; only a small proportion of towns had piped water supply. In the 1980s, the Department of Public Health Engineering (DPHE) started expanding the provision of piped water supply in the urban areas. For the rural areas, DPHE started providing free handpump tubewells in the early 1970s. Initially, the responsibility of operations and maintenance of these tubewells was with the DPHE, but in the 1980s, this responsibility was transferred to the users. With the increase in the capacity of the private sector to manufacture and install tubewells, individuals started installing their own handpump tubewells. Now, almost 80 percent of the total tubewells in the country have been installed by individuals, and these are mainly shallow handpump tubewells for individual households. For the poorer communities, mostly in the rural areas, NGOs provide support with installing tubewells. The tubewells or water points provided by the DPHE and NGOs are shared by more than one household in a community, whereas the tubewells installed by the individuals serve only one household (LGD 2012).

Challenges Facing the Water and Sanitation Sector

The majority of water and sanitation facilities are financed from private sources. Sixty-five percent of the total funding available to the water and sanitation sector is financed by DPs and NGOs and 35 percent by GOB from tariffs and charges (WSS LCG "Open Data" Monitoring Platform). Hanchett et al. (2011) surveyed 53 out of the 481 Union *Parishads* in Bangladesh to assess the long-term sustainability of improved sanitation in Bangladesh. They found that 96 percent of the households paid for their latrines, either from their own funds or with assistance of relatives/friends. Seven percent of the households borrowed money—55 percent from private sources, 40 percent from NGOs, and 5 percent from a cooperative or a bank. For water supply, as indicated in the earlier section, 80 percent of the total tubewells in the country have been installed by individuals from private sources (LGD 2012). Therefore, in order to ensure sustainability of and close monitoring of the sector, the GOB needs to increase its share of financing of these public goods.

Quality of water and sanitation facilities needs improvement. While access has increased substantially, there are still significant challenges if Bangladesh is going to ensure safe, affordable, and sustainable services for all. The quality of sanitation coverage is an emerging area of concern, with over 40 percent of all latrines still classified as being "unimproved" (JMP 2013). According to the Multiple Index Cluster Survey (MICS) carried out by WHO and UNICEF, in 2011 about one-quarter of pit latrines were covered only by a slab, without a water seal, flap, or lid; these latrines are not able to block disease transmission routes. Moreover, one-third of the households share latrines. These latrines connect to a water trap that breaks off soon after it is installed, thus exposing the content of the pit and making the latrine unhygienic (World Bank 2013b). For the poorer segments of the population, access to basic and improved sanitation has increased over the years, but is lagging behind the level of access of the richest quintile of the population (figure 4.3).

For water supply, drinking water is undermined by severe quality issues. For example, 20 percent of the water supply is arsenic contaminated at the source and 12 percent at point of use, and service provision is often unreliable and intermittent (JMP 2013).

Lack of options for better-quality sanitation facilities. Only one type of latrine is available in the rural areas, designed by the DPHE many years back. This design is outdated, of low quality, not user friendly, cumbersome to install, and so forth. In rural areas, these latrines are made by small-scale producers who are not equipped to innovate affordable and durable latrines. Over the years, the cost of

Figure 4.3 Access to Urban Sanitation in Bangladesh by Income Quintile, 1995 and 2008 (%)

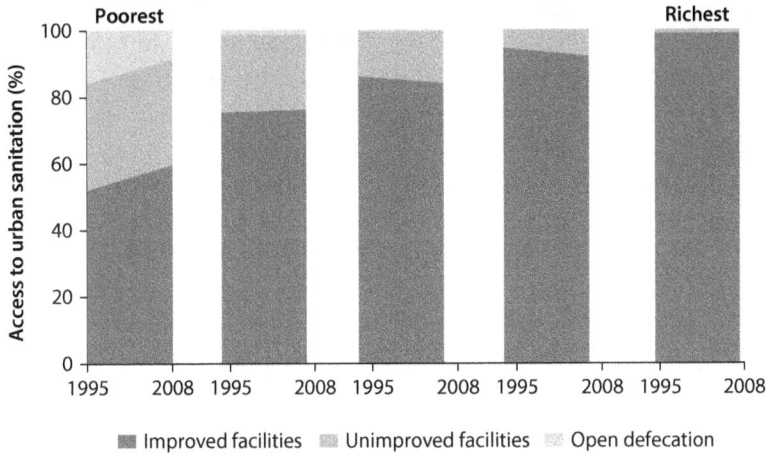

Source: JMP 2013.

the raw materials used for making latrines (such as cement, rod, and brick chips) has gone up considerably, but the prices of the latrines have remained the same. Therefore, the quality of the latrines has declined. For rural areas where there is no running water supply, there is need for user-friendly and affordable latrines with a provision of water. These types of devices have not yet been designed in Bangladesh. Moreover, there is inadequate marketing and promotion of sanitary products. The private sector can play a critical role in investing in research and development in order to offer an array of affordable and user-friendly latrine facilities.

Stronger political commitment needed for further improvement of sanitation facilities. Bangladesh achieved "open-defecation free" status due to a strong political commitment led by the National Sanitation Taskforce coupled with the CLTS initiative. The significant decline in the rates of open defecation, however, cannot be considered as "mission accomplished." The transmission of infection from dirty toilets and unimproved sanitation facilities has to be curtailed in order to reap the full benefits of the investments made thus far in building sanitation facilities. The provision of "improved" sanitation facilities will have to be significantly expanded in order to (i) improve facilities for the 43 percent of the population who are currently using "unimproved" facilities and (ii) reduce the sharing of improved sanitation facilities by more than one household. On average, 27 percent of households share toilets as they do not have their own, and in some areas this number is 50 percent (World Bank 2013b). Hanchett et al. (2011) in their survey of 53 out of the 481 Union *Parishads* in Bangladesh observed that only 44 percent of the sanitation facilities (both improved and unimproved) were clean. They also found that one factor contributing to cleanliness of latrines was ownership of the facility. A latrine owned by a household is more likely to be clean than a shared latrine.

Limited coverage of piped water supply in urban areas. Piped water supply is available only in urban areas. This coverage reaches about 30 percent of the population and only one-third or 102 of the 308 *Pourashavas* (Municipalities). The growing majority of the urban population relies on shallow handpumps connected to each household for drinking water (figure 4.4). With the intermittent and unreliable piped water supply, most households connected to piped water are forced to treat this water for drinking or continue to drink water from handpumps.

Hygiene—the weak link. Lack of hygiene practices has been documented in the SDP 2011–25 as one of the major challenges of the water and sanitation sector (LGD 2012). According to the Bangladesh National Baseline Hygiene Survey 2014, although more than two-thirds of the households had a location near the toilet for postdefecation handwashing, only 40 percent had water and soap available. During a handwashing demonstration, only 13 percent of children aged three to five years and 57 percent of mothers/female caregivers washed both hands with soap. It is believed that these figures are an overestimate of usual practice, as they were observed during a handwashing demonstration. To respond to this challenge, the GOB has formulated a Hygiene Promotion Strategy in 2012 and has emphasized improving hygiene practices in the draft National Strategy for Water Supply and Sanitation 2014.

Hygiene promotion was supposed to be an integral part of the National Sanitation Strategy 2005, which resulted in an increase in the provision of sanitation facilities. However, the 2005 Strategy focused mainly on safe human excreta disposal, which was considered the national priority of that period rather than hygiene. The objective of the 2012 Hygiene Strategy is more ambitious: *"to promote sustainable use of improved water supply and sanitation infrastructures*

Figure 4.4   Access to Urban Water in Bangladesh by Income Quintile, 1995 and 2008 (%)

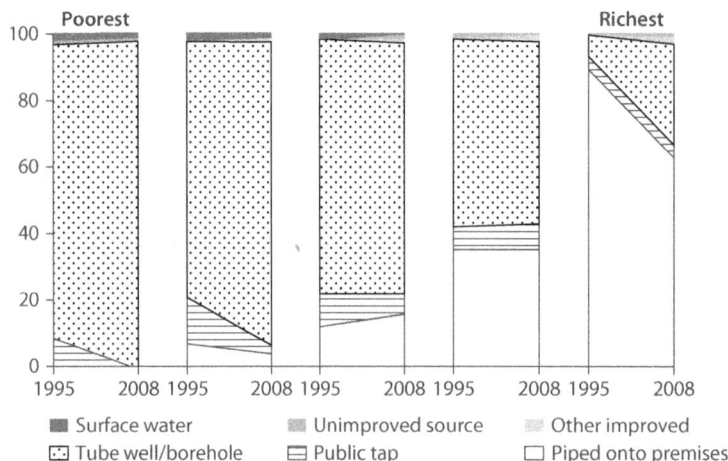

Source: JMP 2013.
Note: The decline in the use of piped water supply in the upper quintile of the urban population reflects the heavy reliance on handpumps for drinking, even when piped water supply connections are present.

and to create an enabling environment ensuring comprehensive hygiene promotion and practices to reduce water and sanitation related diseases." The 2012 Strategy provides a framework for the implementation, coordination, and monitoring of various activities for launching hygiene promotion at national, regional, and local levels. This Strategy has been developed on the basis of a Hygiene Improvement Framework and contains three main components: access to hardware, hygiene promotion, and enabling environment. The Strategy recommends the adoption of an integrated program with all three components for hygiene promotion and delineates the responsibilities of the key ministries, including Ministry of Health and Family Welfare (MOHFW). To highlight the importance of hygiene, the following directions are included in the Strategy for better hygiene:

- Exploring new approaches for hygiene promotion that are effective in translating people's knowledge into practice
- Prioritizing handwashing with soap and menstrual health management
- Addressing specific behavioral domains (such as food hygiene and environmental hygiene)
- Involving key stakeholders (mothers of under-five and school children, health care assistants, and religious and community leaders)
- Collaborating with the private sector for promotion of hygiene-related consumer products like soaps, sanitary napkins, water storage tanks, and washing devices
- Undertaking a national hygiene and sanitation campaign in partnership with the media
- Working with the MOHFW to ensure involvement of health workers, and the MOPME and the Ministry of Secondary and Higher Education (MOSHE) for promotion of hygiene and sanitation practices in schools.

The government entity working in this sector is the DPHE, as detailed in the next subsection. DPHE's niche is in building infrastructure, where progress has been made. DPHE, however, does not have the capacity or the comparative advantage of implementing behavior change communication (BCC activities). There is a need to asses which government agency is better placed to spearhead the task. Perhaps, the MOHFW could take a lead in this area. Nevertheless, clear coordination mechanisms will need to be established to ensure that the "demand" (resulting from successful BCC) is adequately met by "supply" (availability of improved water and sanitation facilities).

Strategies, Institutions, and Interventions in the Water and Sanitation sector

Over the last decade and a half, Bangladesh has formulated numerous policies and strategies to address the challenges faced by the sector. Four legislative acts, two national policies, and five national strategies (excluding the draft 2014 Strategy) have been targeted specifically for the water and sanitation; they are described below. Additionally, there are various policies and plans for the sector. These documents provide strategic directions and broad frameworks for developing action plans and interventions in the water and sanitation sector. The strategies and policies are comprehensive and cover almost all of the issues highlighted in the preceding sections. However, these strategies and policies need to be translated into action, which remains a challenge.

Legal framework: The Water Act 2013 of the GOB provides a legal framework for the sector, along with the Water Supply and Sewerage Authority (WASA) Act 1996, Environmental Conservation Act 1995, Environmental Conservation Rules 1997, and the different Local Government Acts 2009 for the City Corporations, the *Pourashavas* (Municipalities), the Upazila *Parishads*, and the Union *Parishads*.

National level policies: The Perspective Plan (2010–21) of the GOB prioritizes interventions for ensuring access to drinking water, sanitation, and good hygiene practices for all. The GOB has also submitted to the United Nations (UN) its post-2015 development agenda (2016–30), with the goal of "Safe and sustainable sanitation, hygiene and drinking water used by all." UN-Water (the United Nations interagency mechanism for all freshwater and sanitation related matters) has proposed specific targets and indicators to meet Bangladesh's post-2015 Sustainable Development Goals.

Sector strategies: The SDP 2011–25 highlighted the need for having an integrated strategy for the water and sanitation sector (LGD 2012). The GOB has formulated a series of policies and strategies over the years without a clear plan of harmonization and integration of national efforts. The SDP 2011–25 documents how national policies address the main areas of concern—reduction of open defecation, ensuring supply of safe water, mitigation of arsenic contamination, and forging strategic partnerships for expanding coverage of water and sanitation facilities. However, there are gaps and overlaps among these different strategies. The GOB has, therefore, drafted the National Water Supply and Sanitation Strategy 2014, in an effort to integrate the various strategies (listed below) and provide one coherent sector strategy covering issues of water, sanitation, and hygiene.

The goal of the draft 2014 Strategy is to ensure *"safe and sustainable water supply, sanitation, and hygiene services for all, leading to better health and well-being."* With this overall goal, a set of 17 strategies has been formulated, broadly grouped into three themes: increasing WASH interventions, addressing emerging challenges, and strengthening sector governance (table 4.5). The five-year draft Strategy intends to provide uniform strategic guideline to the key stakeholders of the sector, including government institutions, the private sector, and NGOs. The draft 2014 Strategy delineates the roles and responsibilities of the other ministries, including MOHFW and the Ministry of Primary and Mass Education (MOPME).

Table 4.5 Priority Areas of the Draft National Water Supply and Sanitation Strategy 2014

Thematic area	Specific strategy
WASH interventions	1. Ensure safe drinking water
	2. Give priority to arsenic mitigation
	3. Undertake specific approaches for hard-to-reach areas and vulnerable people
	4. Move ahead on the sanitation ladder
	5. Establish fecal sludge management
	6. Manage solid waste judiciously
	7. Improve hygiene promotion
	8. Mainstream gender
	9. Facilitate private sector participation
Emerging challenges	10. Adopt integrated water resource management
	11. Address growing pace of urbanization
	12. Cope with disaster, adapt to climate change, and safeguard the environment
	13. Institutionalize research and development
Sector governance	14. Undertake integrated and accountable development approach
	15. Recover cost of services while keeping a safety net for the poor
	16. Strengthen and reposition institutions
	17. Enhance coordination and monitoring

Source: LGD 2014b

Prior to the formulation of the draft 2014 Strategy, the main policies of GOB specific to the water and sanitation sector were as follows:

- National Policy for Safe Water Supply and Sanitation 1998—stipulated that the government's goal is to ensure that all people have access to safe water and sanitation services at an affordable cost. The policy emphasized elements of behavioral change and sustainability through user participation in planning, implementation, management, and cost sharing.
- National Water Policy 1999—provided guidance on management of the country's water resources by all the concerned ministries, agencies, departments, and local bodies that are assigned responsibilities for the development, maintenance, and delivery of water and water related services as well as the private users and developers of water resources. This policy attached special importance to the conjunctive use of ground and surface water.
- National Policy for Arsenic Mitigation and Implementation Plan 2004—gives "preference to surface water over groundwater." The policy provides a guideline for mitigating the effect of arsenic on people and the environment

in a realistic and sustainable way. At the operational level, it provided a conceptual shift from single use of water (such as through handpumps) for drinking water and motorized deep tubewells for irrigation, to multiple use of water from deep tubewells.

- National Sanitation Strategy 2005—sets out the goal to achieve 100 percent sanitation by 2010 and identified six areas of concern to be addressed (open defecation, hardcore poor remaining underserved, use of unhygienic latrines, lack of hygiene practices, urban sanitation, and solid waste and household wastewater disposal not fully addressed).
- Pro-Poor Strategy for Water and Sanitation Sector 2005—provides the operational definition of hardcore poor households, basic minimum service standards, and mechanisms for targeting and organizing the households and for administering subsidies.
- National Cost Sharing Strategy for Water Supply and Sanitation in Bangladesh 2011—proposes ways to standardize water supply and sanitation services and recommends cost sharing modalities to ensure affordable, equitable, and sustainable water and sanitation services for all. The Strategy includes an analysis of the sectoral issues, the needed policy reforms and institutional development, as well as economic pricing of the services.
- National Hygiene Promotion Strategy for Water Supply and Sanitation in Bangladesh 2012—provides a framework for the implementation, coordination, and monitoring of various activities for launching hygiene promotion at national, regional, and local levels.
- National Strategy for Hard to Reach Areas and People of Bangladesh 2012—provides definitions and criteria to be used to identify hard-to-reach areas and people, and documents the challenges in reaching out to them. The Strategy provides an outline of activities that need to be carried out in order to address the problems faced by the different types of hard-to-reach areas (like coastal areas and wetlands).

Appendix A provides details of some of the key strategies and policies of the GOB.

Institutions and Water Supply and Sanitation Interventions

The Local Government Department (LGD) of the Ministry of Local Government Rural Development and Cooperatives (MOLGRD&C) is responsible for the overall development of the water and sanitation sector, while the DPHE and the WASAs function under the administrative control of LGD. In 1983, WASAs were established in Dhaka and Chittagong cities as special-purpose institutions, being responsible for water supply, sewerage, and drainage in those areas. In 2008, Khulna WASA was created. The DPHE is responsible for implementation of water supply and sanitation interventions in the public sector in both rural and urban areas (with the exception of the areas covered by the WASAs). In urban areas, DPHE was originally responsible for the water supply and sanitation services, but gradually the *Paurashavas* and the City Corporations are becoming more involved in planning, implementation, and management of the water systems. In addition to DPHE, the Local Government Engineering Department (LGED), under the LGD, implements water supply and drainage services in the urban areas.

The GOB started its initial intervention in the water supply and sanitation sector with the objective of developing an effective service delivery mechanism. After Bangladesh's independence in 1971, emphasis was placed on rehabilitation of the damaged water supply and sanitation facilities and expansion of coverage of services by DPHE. These services were provided mostly free of charge. Users did not play a role in decision making, cost sharing, and operations and maintenance. Over the years, user participation has increased significantly and rural communities are now responsible for operation and maintenance of handpump tubewells and receive training for the purpose.

Most of the *Paurashavas* and the Union *Parishads* have Water Supply and Sanitation Committees (WATSAN), which are composed of the user communities for supervising water and sanitation related activities. In rural areas, the Local Government Institutions (LGIs) ensure coordination of the water supply and sanitation services that are provided; while the WATSAN Committees of the Upazila *Parishads* and Union *Parishads* have the responsibility to coordinate the activities of DPHE, NGOs, and other stakeholders. A National Forum for Water Supply and Sanitation (NFWSS) has been established in the LGD for ensuring coordination at the national level between the key stakeholders, including government agencies, NGOs, DPs, and the private sector. Since the time Bangladesh achieved open defecation–free status, only one-third of the WATSAN Committees are functional, mostly in the areas where there is NGO support.

Almost all of the interventions in the sector have focused on building infrastructure for providing access to water and sanitation facilities. As a result, Bangladesh has been able to move from traditional drinking water sources (ponds, dug wells, and rivers) to handpump tubewells. In sanitation, the country has transitioned from open defecation to "fixed-point defecation" as detailed out in the following section. The "softer" sides of the sector (quality, hygiene, innovation, and so forth) have not progressed as much.

A list of the major projects (not an exhaustive list) is included in appendix B. Some of the main interventions of the government include the Bangladesh Water Supply Programme Project (2004 to 2009), Rural Water Supply Project throughout the country (fifth phase, 2004 to 2010), National Sanitation Project (second Phase, 2007 to 2011), Water Management Improvement Project (2007 to 2015), Chittagong Water Supply Improvement and Sanitation Project (2010 to 2015), Dhaka Water Supply and Sanitation Project (2009 to 2015), and the Bangladesh Rural Water Supply and Sanitation Project (2012 to 2017). All of these are focused on improving the coverage of water supply and sanitation facilities.

Opportunities for Improving Nutrition through Interventions in Water and Sanitation

Table 4.6 presents a theoretical framework developed by Dr. Mduduzi Mbuya (Sanitation Hygiene, Infant Nutrition Efficacy Study, Zimbabwe) to identify key factors and potential areas of intervention in reducing fecal-oral contamination; it has been adapted for this report based on discussions with the authors. The framework is essentially an operationalization of concepts that have been described in the literature (Curtis, Cairncross, and Yonlim 2000) and critical pathways of fecal-oral contamination, identified in recent studies (Ngure et al. 2013 and Ngure et al. 2014). While it is still being tested in clinical trials, the observational studies that led to its development render it a plausible framework. Based on the F-diagram (that forms the basis for most WASH programs), it presents an intersection between nutrition, WASH, and early childhood development, and also introduces the concept of "child-specific WASH."

The framework helps program planners, implementers, and evaluators ascertain the adequacy/completeness of a package of WASH interventions for mitigating fecal-oral contamination among infants and young children.

Reducing fecal load in the living environment. The use of quality sanitation facilities is central to reducing exposure to human fecal matter. The GOB has made significant progress in the construction and provision of sanitation facilities for its citizens as detailed in the previous section. However, improvements on the quality (preferably those that facilitate and ensure fly control) and utilization of facilities require more work if indeed fecal-oral contamination is to be curtailed. Innovative sanitation marketing, using a range of interventions to raise householder's demand for improved sanitation and creating an enabling policy and institutional environment, will need to be pursued to improve the hygiene conditions and reduce fecal load in the living environment.

Reducing fecal contamination via hands. Handwashing by all household members (including children) after key potential contamination events, such as after fecal contact, before handling food, and before child feeding is crucial for reducing fecal-oral contamination. In most cases, water is available inside the latrines (either carried in mugs by people before entering the latrine, or stored in jars, or piped onto the latrine), which is used by people for cleaning themselves. However, handwashing after using the latrines remains low due to the absence of handwashing stations, as well as soap/hand cleansing substances inside or outside the latrines. Even where water and soap are available inside/near the latrine, handwashing remains low as documented in the National Baseline Hygiene Survey 2014. Emotional/social drivers can bring about a change in practice. Disgust is one such negative emotional/social driver that affects people, while the sense of pride that comes with using improved latrine is a positive driver (Hanchett et al. 2011). The "Open Defecation Free" campaign has successfully raised knowledge among the rural population in Bangladesh about how open defecation pollutes the environment and contaminates food and water sources. This, along with religious and cultural values of "purity" and "pollution" around

Table 4.6 Framework for a Package of WASH-related Interventions

Intervention objective	Hardware—inputs		Software—behavior change messages	
	Access (provision)	Practical/technical considerations	Utilization (encouragement, demand creation)	Triggers/motivators
Reduce fecal load in living environment	Household sanitary facility (toilet) including tools for the safe collection, transport, and disposal of child feces (potties, scoopers, and so on).	Preferably one that facilitates or ensures fly control.	Use of sanitary facilities by all household members. Safe disposal of child feces.	Disgust has been shown to be an effective trigger for behavior change (Hanchett et al. 2011).
Reduce fecal transmission via hands	Handwashing facility, soap/ cleansing substance, water (quantity).	Placement of the handwashing facility can be an important (visual) cue to behavior— in close proximity to the sanitation facility. Soap or other cleansing substances (like ash) should be available along with water to ensure effective handwashing.	Handwashing by all household members (including children) after key potential contamination events (after fecal contact, before handling food, before child feeding, and so on).	Disgust, nurture, and comfort are effective in triggering handwashing (Republic of Kenya, UNICEF, World Bank 2007).
Improvement of drinking water quality	Safe water source. Drinking water storage containers. Treatment agent/model (for example, solar, chlorine) to ensure water quality at the point of use.	Water treatment agent should meet organoleptic (taste, smell) expectations of household members.	Water treatment at the point of use. Drinking of treated water by all household members.	Associated taste and smell of treated water with cleanliness. Nurture is an effective motivator (Republic of Kenya, UNICEF, World Bank 2007) for promoting provision of treated water to children.
Avoidance of child fecal ingestion during mouthing and exploratory play (for example, geophagy [the practice of eating earthy substances], consumption of chicken feces.	Protective child play space.	The play space should ensure that the child is protected from environmental contamination while ensuring their developmental needs for exploration are met.	Awareness of risks associated with playing in a contaminated environment, for example geophagy, direct/indirect consumption of animal feces.	Risk awareness. Nurture.

41

feces, may have contributed to sustained use of latrines (Hanchett et al. 2011). The media can play a strong role in promoting handwashing messages. Data (univariate and multivariate analyses) from a nationwide cross-sectional survey in 800 households in Kenya indicate that water access, levels of education, media exposure, and media ownership are strongly associated with handwashing with soap (Schmidt et al. 2009). BCC, similar to these but customized for promotion of handwashing, will need to be pursued.

Improvement of drinking water quality. In settings that are not served by reliable water treatment and distribution systems, diarrheal disease is usually prevalent in the population at all times. In such settings, health benefits derived from improvements in the quality of drinking water can be immense. It is the micro-bial quality of water at the point of use, rather than source, which has been shown to be more effective in reducing the occurrence of endemic diarrhea (WHO 2002). For example, a household might have access to clean water from a well, borehole, or a community source; but if that water needs to be trans-ported to the household and stored before use, then there is risk for microbial contamination. In this regard, household water treatment and safe storage inter-ventions (such as boiling, solar treatment, ultraviolet disinfection with lamps, and filtration) could be followed by chlorination and storage in a protected or improved vessel (WHO 2002). These steps can lead to dramatic improvements in drinking water quality and reductions in diarrheal disease. As noted earlier, in the absence of a formal water treatment facility, households in Bangladesh are forced to treat piped water to make it drinkable or use handpumps.

Avoidance of child fecal ingestion during mouthing and exploratory play. In rural and/or slum communities, young children crawl and play in environments where they may come into contact with soil that maybe contaminated with both human and animal feces. Young children regularly mouth objects as part of nor-mal development. Ngure et al. (2013) noted that "active exploratory ingestion of soil and chicken feces have the greatest risk of fecal bacteria exposure in terms of high microbial load." As such, despite adequate access to clean water and appropriate sanitation facilities, crawling on contaminated soils and kitchen floors can expose infants and young children to low, but frequent, dosages of fecal bacteria, which can elicit physiological responses of tropical/environmental enteropathy. Interventions aimed at preventing children's exposure to fecal bac-teria are critical, if not more important than handwashing and water treatment, as these transmission pathways may lead to greater dosages of exposure (Ngure et al. 2013). Therefore, the traditional approach of WASH interventions (water supply and construction of sanitation facilities) is unlikely to prevent fecal-oral contamination among young children without addressing children-specific risk factors in the government programs.

Overall, the GOB policies and strategic direction in the water and sanitation sector are comprehensive and cover the key factors that could have a signifi-cant impact on nutrition. The challenge, however, is the translation of these policies and strategies into action. Thus far, the programmatic focus has been on the provision of water and sanitation facilities and little on implementing the strategies related to BCC and other approaches to promote appropriate hygiene conditions and practices. In addition, behavior change for the uptake of handwashing and a focus on reducing animal fecal contamination are areas that require attention in the GOB's interventions.

CHAPTER 5

Conclusions

Recommendations

The discussions in the preceding chapters have pointed out the following needs related to malnutrition: (i) improve quality of water supply and sanitation facilities; (ii) strengthen the Health, Nutrition, and Population (HNP) interventions to address undernutrition; (iii) implement activities for promotion of hygiene; and (iv) ensure adequacy of health care, food, and environmental health, without which the problem of malnutrition cannot be addressed.

Against this backdrop, the following recommendations are made which may be considered by the Government of Bangladesh (GOB). The first two sets of recommendations are specific for the water and sanitation and the HNP sectors, while the third set of recommendations is overarching and multisectoral.

Making Water and Sanitation Activities More "Nutrition-sensitive"

1. **Improve quality of water and sanitation facilities.**

It is critical to improve the quality of water (at source, in storage, as well as at point of consumption) and sanitation facilities to limit transmission of infection. There is also a need to ensure that all members of the households that have piped water supply also have safe drinking water. Awareness-raising campaigns along with emotional/social drivers can be effective in ensuring attention to the water supply. In the rural areas of Bangladesh, where the majority of the population get water from nonpiped sources, it is necessary to initiate practices for safe collection and transportation of water from the community points and then purification and safe storage of water. For the urban areas, where a piped water supply is provided to households, the water needs to be treated to make it drinkable; thus, it is important to promote safe water storage facilities.

When devising mechanisms for improving coverage and quality of sanitation facilities[1] in Bangladesh, the following lessons should be considered (Hanchett et al. 2011):

- Female-headed households are more likely to have an improved or shared latrine compared to households headed by males. A possible explanation is related to the concept of *purdah*[2] that exists in Muslim and Hindu cultures. A latrine offers women privacy for defecating, urination, and menstruation management, which allows them to adhere to *purdah* and avoid the shame of being seen by men at these times.
- Access to private sector providers is a factor that enables sustained use of improved latrines (with provision for flushing the feces and availability of water).
- Social norms around open defecation and latrine use have positively changed, which likely was a result from sanitation and hygiene promotion. One plausible contributor to this shift in social norms is that the Behavior Change Communication (BCC) campaign directed toward households was fairly pervasive: campaign messages were communicated through various channels and settings, including messaging by Union *Parishad* members or officers at meetings, rallies, over loudspeaker announcements, and household visits by Union *Parishad* members or nongovernmental organization (NGO) workers.
- Factors correlated with unsustained use of improved latrines are as follows:
  o poverty is a factor that affects the sustained use of latrines
  o Severe natural disasters have an effect on sustained use of latrines
  o lack of local leadership may affect sustained use of latrines

2. Strengthen implementation of hygiene-related activities.

Hygiene remains the weakest link in the water and sanitation sector. At the strategic level, the 2014 draft National Water Supply and Sanitation Strategy adequately addresses this issue. It is now, therefore, critical to finalize the draft 2014 Strategy and implement the action plan. Moreover, the GOB will need to monitor progress of the implementation of the action plan through a high-level intersectoral committee to ensure better coordination between the various ministries. Particular emphasis will need to be placed on increasing the availability of handwashing stations, soap/hand cleansing substances, and water at the sanitation facilities, and ensuring that these are used. Also, mass-media campaigns should be undertaken for promotion of handwashing. Various ministries, including health, education, local government, and water resources, have a key role to play in promoting handwashing practices. Perhaps the Ministry of Health and Family Welfare (MOHFW) could take a lead role in implementing the BCC while the Department of Public Health and Engineering (DPHE) ensures the supply-side availability of improved water and sanitation facilities.

Improvements in the Nutrition Activities of the Health Sector

3. **Strengthen the effectiveness of the National Nutrition Services (NNS).**

The MOHFW needs to define a prioritized set of activities that are critical for improving undernutrition, particularly improved hygiene practices. As the recent NNS assessment indicates that the current delivery platform is not being effective, alternative service delivery mechanisms will need to be explored, including the involvement of NGOs, to extend outreach and achieve greater targeted coverage. NNS, due to its modality of service delivery through public health facilities, is targeted toward mostly the poorer and disadvantaged population, as the richer households tend to opt for private health care services. The MOHFW, therefore, will continue to underserve a large segment of the population with high undernutrition rates if delivery continues through public health facilities. The government should actively consider engaging the media and the private sector for the required BCC as well as promoting handwashing through health sector interventions.

4. **Consider the preventive aspects of nutrition, rather than just treatment.**

At present, under NNS small corners for integrated management of childhood illnesses and nutrition (titled integrated management of childhood illnesses and nutrition [IMCI&N] corners) are being set up at the health facilities. This modality has the disadvantage of only covering sick children by the nutrition services. The MOHFW needs to transition from the IMCI&N corners as a central delivery platform for the NNS to investing more deeply in an alternative and predominantly outreach-based platform for delivering core services to households and children. These might be called "well-child spots" and located near or at the existing health facilities at upazila levels and below (World Bank 2014). Since tropical/environmental enteropathy is a subclinical condition, it is important to advocate for "prevention" of undernutrition rather than "treatment."

Multisectoral Response to Undernutrition

5. **Strengthen the health sector response, but also build a nonhealth, multisectoral response for addressing undernutrition.**

The determinants of undernutrition are multisectoral, yet attempts to implement multisectoral programs have proved largely unsuccessful. Multisectoral nutrition planning agencies have been stymied by the limited control they have over different sectors' resource allocation processes, while sectorally defined priorities have hindered collaboration between sectors. A more realistic response is to "plan multisectorally, implement sectorally" (Maxwell and Conway 2000). Operationally, this involves identifying interventions within sectors that have the potential to significantly improve nutrition and mobilizing resources specific to that sector. An important factor for this strategy to work is to ensure sensitization of sector-specific stakeholders. They need to understand how their sectoral outcomes can be improved through addressing undernutrition and how their sector can contribute to achieving this. The starting point for the approach will be to draw upon the existing policies and strategies and identifying potential avenues for making the sector-specific interventions more "nutrition-sensitive" (World Bank, DFID, Government of Japan, and Rapid Social Response 2013a).

6. **Synchronize the efforts the various sectors to align with the overall goal of reducing undernutrition.**

Individual efforts by MOHFW and other ministries have the desired impact on undernutrition rates. The relevant sectors—HNP, water and sanitation, education, local government, and agriculture—need to come forward and integrate their efforts to attain the broader national goal of improving nutritional outcomes. To enable this coordination, it is necessary to ensure that alleviating undernutrition remains a high-level policy priority. Undernutrition and poverty are interrelated. In order to augment economic development, it is crucial to ensure that undernutrition is addressed.

There are challenges in coordinating efforts of the HNP sector with those of the water and sanitation sector. Lessons from successful collaboration could be drawn upon to mitigate the challenges. In the Amparae district of Sri Lanka, for example, the introduction of student brigades had a significant impact on hygiene behavior change, contributing to adequate utilization of the water, sanitation, and hygiene (WASH) facilities available in the schools (Yael et al. 2014). Active involvement of the health professionals in WASH is crucial for accelerating and consolidating progress relating to the health and nutrition outcomes (Bartram and Cairncross 2010). The HNP sector needs to take a more assertive role in promoting use of WASH facilities. The sector could include WASH as one of the essential components of all HNP policies, monitor handwashing practices using routine health surveillance, and use WASH messages in advocacy and outreach (Cairncross et al. 2010) through HNP sector interventions.

Notes

1. The Joint Monitoring Programme (JMP) defines improved sanitation facility as use of flush to piped sewer system/septic tank/pit latrine; ventilated improved pit latrine; pit latrine with slab; and/or composting toilet.

2. *Purdah* is a state of seclusion and/or privacy, particularly from men, emanating from religious and/or societal beliefs and applicable for lifestyle.

Glossary

Anemia — Low level of hemoglobin in the blood, as evidenced by a reduced quality or quantity of red blood cells; 50 percent of anemia worldwide is caused by iron deficiency.

Body mass index (BMI) — Body weight in kilograms divided by height in meters squared (kg/m^2). This is used as an index of "fatness" among adults. Both high BMI (overweight, BMI greater than 25) and low BMI (thinness, BMI less than 18.5) are considered inadequate.

Catabolism — Set of metabolic activities that break down larger molecules into smaller units.

Cochrane Review — Cochrane Reviews are systematic reviews of primary research in human health care and health policy and are internationally recognized as the highest standard in evidence-based health care. They investigate the effects of interventions for prevention, treatment, and rehabilitation. They also assess the accuracy of a diagnostic test for a given condition in a specific patient group and setting. These results are published online in *The Cochrane Library* (http://community.cochrane.org/cochrane-reviews).

Low birthweight — Birthweight less than 2,500 grams.

Malnutrition — Various forms of poor nutrition caused by a complex array of factors including dietary inadequacy, infections, and sociocultural factors. Both underweight or stunting and overweight are forms of malnutrition.

Obesity — Excessive body fat content; commonly measured by BMI. The international reference for classifying an individual as obese is a BMI greater than 30.

Overweight	Excess weight relative to height; commonly measured by BMI among adults (see above). The international reference is as follows: • –29.99 for grade I (overweight) • –39.99 for grade II (obese) • > 40 for grade III. For children, overweight is measured as weight-for-height z-scores of more than two standard deviations above the international reference.
Stunting (measured as height-for-age)	Failure to reach linear growth potential because of inadequate nutrition or poor health. Measured as height-for-age z-scores that are more than two standard deviations below the median value of the reference group. Usually a good indicator of long-term undernutrition among young children.
Union *Parishad*	The oldest and lowest tier of local government representing 10–15 villages with around 5,000 households. Each union is composed of 13 elected representatives including a chair, nine members (one from each ward), and three women members who are elected to reserved seats.
Vitamin A deficiency	Tissue concentrations of vitamin A low enough to have adverse health consequences such as increased morbidity and mortality, poor reproductive health, and slowed growth and development, even if there is no clinical deficiency.
Wasting (measured by weight-for-height)	Weight divided by height that is two z-scores below the international reference. It describes a recent or current severe process leading to significant weight loss, usually a consequence of acute starvation or severe disease. Commonly used as an indicator of undernutrition among children, and especially useful in emergency situations such as famine.
WASH intervention	Improved water quantity and quality, sanitation, and hygiene.
z-score	The deviation of an individual's value from the median value of a reference population, divided by the standard deviation of the reference population.

Policies and Strategies in the Water and Sanitation Sector

National Policy for Safe Water Supply and Sanitation (1998)

In 1998, the Local Government Division (LGD) of the Ministry of Local Government, Rural Development and Cooperatives (MOLGRD&C) prepared the National Policy for Safe Water Supply and Sanitation. The Policy stipulates the Government of Bangladesh (GOB) has the goal to ensure that all people have access to safe water and sanitation services at an affordable cost. The objectives of the National Policy for Safe Water Supply and Sanitation 1998 were *"to improve the standard of public health and to ensure improved environment."* The policy recognized *"that physical provision of services alone is not a sufficient precondition for sustainability or improvement of health and wellbeing of the people."* It therefore emphasized elements of behavioral change and sustainability through user participation in planning, implementation, management, and cost sharing. The following specific directions are included for achieving the objectives:

- Facilitating access of all citizens to basic level of services in water supply and sanitation
- Bringing about behavioral changes regarding use of water and sanitation
- Reducing incidence of waterborne diseases
- Building capacity in local governments and communities to deal effectively with problems relating to water supply and sanitation
- Promoting sustainable water and sanitation services
- Ensuring proper storage, management, and use of surface water and preventing its contamination
- Taking necessary measures for storage and use of rainwater
- Ensuring storm-water drainage in urban areas.

The 1998 Policy also recognized the need for transitioning from the traditional service delivery mechanism and institutionalizing strategic partnerships between government and nongovernment organizations (NGOs) and civil society organizations.

National Water Policy (1999)

Around the same time of the National Policy for Safe Water Supply and Sanitation, the GOB developed the National Water Policy 1999 spearheaded by the Ministry of Water Resources. The Policy documented the main challenges facing water resources management in Bangladesh at that time: seasonal food and water scarcity, additional water needs of an expanding population, and issues related to river bank erosion and sedimentation in rivers that affect navigability. The Policy emphasized the need for a total water quality management (to address issues of salinity and deteriorating quality of surface and groundwater, as well as pollution) and maintenance of the ecosystem. Other issues documented included meeting water requirements of multiple sectors with limited resources, efficient and socially responsible use of water use, identifying roles of the public and private sectors, and decentralization of state activities. The Water Policy 1999 provides broad principles for effective development of water resources and ensuring its rational utilization. Its objectives were as follows:

- Address issues related to the harnessing and development of all forms of surface and groundwater and an efficient and equitable management of these resources
- Ensure the availability of water for all, particularly the poor and the underprivileged, and take into account the particular needs of women and children
- Accelerate the development of sustainable public and private water delivery systems with appropriate legal and financial measures and incentives, including delineation of water rights and water pricing
- Bring institutional changes that will help decentralize the management of water resources and enhance the role of women in water management
- Develop a legal and regulatory environment that will help the process of decentralization and sound environmental management, and improve the investment climate for the private sector in water development and management
- Develop knowledge and capability that will enable the country to design future water resources management plans by itself with economic efficiency, gender equity, social justice, and environmental awareness to facilitate achievement of the water management objectives through broad public participation.

National Policy for Arsenic Mitigation and Implementation (2004)

To address the problem of arsenic contamination in groundwater, the GOB formulated the National Policy for Arsenic Mitigation and Implementation in 2004. The goal of the Policy is to ensure that safe water, free from arsenic, is made available for drinking, cooking, and agricultural purposes and that all cases of arsenicosis are diagnosed and managed. It provides guidance for mitigating the effect of arsenic on people and environment in a realistic and sustainable way. At the operational level, it provides a conceptual shift from single use of water such as through handpumps for drinking water and motorized deep tubewells for irrigation, to multiple use of water from deep tubewells (as tubewells are widely used in Bangladesh). The Policy attaches "preference to surface water over groundwater as the source of water supply." The Policy does not cover issues relating to cleanliness of water.

The National Water Management Plan (2004)

The Water Resources Planning Organization (WARPO) of the Ministry of Water Resources formulated the National Water Management Plan 2004. The objective of the Plan is to provide necessary advice on follow-up actions to be taken for implementing the National Water Policy 1999. It sets out specific steps to *ensure development of effective institutions and legal and regulatory measures and to enable efficient and equitable management of the sector as a whole.* The Plan provides time-bound activities with well-defined targets for development of the main rivers; water and sanitation facilities for towns and rural areas and four major cities (Dhaka, Chittagong, Khulna, and Rajshahi); disaster management; agriculture and water management; and natural environment and aquatic resources. For the towns and rural areas, the principal objectives are to provide a safe and reliable supply of potable water and sanitation services to all the inhabitants in the towns and rural areas, along with effective facilities for wastewater disposal to safeguard public health and protect the environment. And in selected towns with facilities of economic importance, flood protection is prioritized while a phased implementation plan is proposed for ensuring reasonable flood protection facilities for other districts and upazilas. Similarly for the four major cities, the overall objectives are "provision of effective facilities to safeguard public health and the environment; attainment of significantly improved standards of operational efficiency and service provision with active community participation and consultation; promotion of private sector participation in water supply and sanitation; and provision of affordable and sustainable services to all city dwellers with particular emphasis on the poor and disadvantaged."

National Sanitation Strategy (2005)

In 2005, the GOB embarked on the goal to achieve 100 percent sanitation by 2010 as documented in the National Sanitation Strategy 2005 of the LGD, MOLGRD&C. In 2009, the goal was revised to "100 percent sanitation by 2013." The main objective of the Strategy was to identify *"ways and means of achieving the national goal through providing uniform guidelines for all concerned."* It identifies six areas of concern to be tackled: open defecation; hard-core poor remaining underserved; use of unhygienic latrines; lack of hygiene practices; urban sanitation; and solid waste and household wastewater disposal not fully addressed. It documented that the "use of safe drinking water and sanitation facilities, together with improved hygiene practices, has a direct impact on poverty by reducing the vulnerability of the poor people." It is clearly spelled out that the focus of the Strategy was to address issues related only to unhygienic defecation. The Strategy specified the definition of 100 percent sanitation as no open defecation, hygienic latrines available for all, use of hygienic latrines by all, proper maintenance of latrines, and improved hygienic practices. Also, the Strategy defined the basic minimum level of sanitation service as every household having access to a safe hygienic latrine, either a separate household latrine, shared latrine by two households, or a community latrine. The importance of coordination and formation of strategic partnerships with NGOs and civil society organizations is emphasized.

National Sector Development Programme for Water Supply and Sanitation 2005

A 10-year National Sector Development Programme for Water Supply and Sanitation 2005 (SDP 2005) was developed by the LGD of MOLGRD&C, which served as a planning document for the GOB and development partners (DPs). This has been updated with the SDP 2011–25 to focus more on hygiene promotion and incorporate needs of the lagging areas (LGD 2012). The SDP 2011–25 points to the need for having an integrated strategy for the water and sanitation sector. As evident from the preceding paragraphs, the GOB formulated a series of policies and strategies without a clear plan of harmonization and integration of national efforts. The SDP 2011–25 documents that while the national policies cover the main areas of concerns—reduction of open defecation, ensuring supply of safe water, mitigation of arsenic contamination, and forging strategic partnerships for expanding coverage of water and sanitation facilities—there are gaps and overlaps among these different strategies. The major challenges documented by the SDP 2011–25 include, among others, the growing pace of urbanization; inadequate and inappropriate urban sanitation; limited piped water supply coverage in the urban areas; and hygiene being a weak link in the sector. The SDP 2011–25 proposes that all the existing strategies be streamlined and a single strategy be prepared covering the known and emerging sector issues. The

LGD of MOLGRD&C in consultation with various stakeholders has, therefore, drafted the National Strategy for Water Supply and Sanitation 2014 to address all of these issues including promotion of hygiene practices.

National Cost Sharing Strategy for Water and Sanitation for Hard-to-Reach Areas of Bangladesh 2011

The goal is *"to provide functional ways and means for water supply and sanitation in Bangladesh to facilitate standardization of and increased access to water supply and sanitation services to all by 2025, and to make services affordable, equitable, and sustainable, at cost."* The Strategy proposes to gradually increase the cost sharing of the consumers and decrease subsidies. Specific goals of the Strategy are (i) recovering costs of services; (ii) gaining financial self-sufficiency; (iii) standardization of water supply and sanitation services; and (iv) ensuring sustainability. Additionally, the Strategy emphasizes on increasing citizens' responsibility and ownership of the facilities.

National Strategy for Water and Sanitation for Hard-to-Reach Areas of Bangladesh 2012

The primary objective is *"to improve safe drinking water and sanitation coverage in hydro-geologically and socio-economically difficult areas where people have services much less than the national standard."* The Strategy provides six measurable and usable indicators to identify hard-to-reach areas and then a separate set of criteria to classify these areas around three categories. Challenges faced by the specific categories are highlighted in the Strategy along with a set of policy recommendations to address these challenges.

National Hygiene Promotion Strategy for Water Supply and Sanitation in Bangladesh 2012

The objective of the 2012 Hygiene Strategy is *"to promote sustainable use of improved water supply and sanitation infrastructures and to create an enabling environment ensuring comprehensive hygiene promotion and practices to reduce water and sanitation related diseases."* The 2012 Strategy provides a framework for the implementation, coordination, and monitoring of various activities for launching hygiene promotion at national, regional, and local levels. This Strategy has been developed on the basis of a Hygiene Improvement Framework which has three main components: access to hardware, hygiene promotion, and enabling environment. The Strategy recommends the adoption of an integrated program with all three components for hygiene promotion and delineates the responsibilities of the key ministries including the Ministry of Health and Family Welfare (MOHFW).

National Strategy for Water Supply and Sanitation 2014 (Draft)

The goal of this draft Strategy is to ensure *"safe and sustainable water supply, sanitation and hygiene services for all, leading to better health and well-being."* With this overall goal, a set of 17 strategies has been formulated broadly grouped into three themes: increasing water, sanitation, and hygiene (WASH) interventions, addressing emerging challenges, and strengthening sector governance. The five-year draft Strategy intends to provide uniform strategic guidelines to the key stakeholders of the sector, including government institutions, the private sector, and NGOs. The draft 2014 Strategy delineates the roles and responsibilities of the other ministries, including MOHFW and the Ministry of Primary and Mass Education (MOPME).

Interventions in the Water and Sanitation Sector

Some of the projects in the water and sanitation sectors are summarized below.

Sl.	Title	Period	Cost (Tk million)	Objective
1	IDB-assisted water supply facilities in the coastal belt of Bangladesh (second Phase)	July 2003 to December 2008	GOB: 76.722 PA: 496.8 RPA: Tk 496.2 million	i) Make safe water available to the inhabitants of project area ii) Enhance the use of safe water to reduce the occurrence of waterborne diseases iii) Ensure arsenic free water
2	Bangladesh Water Supply Project	July 2004 to June 2009	GOB: 3,198.20, PA: 2,320 RPA: Tk 1,490 million	Contribute to Bangladesh's efforts to achieve the Millennium Development Goals in water supply and sanitation by 2015. Specially, the project will pilot innovative measures to scale up the provision of safe water supply free from arsenic and pathogens in rural and small towns. This will take place by the following: i) Promoting rural piped water supply with private sector participation ii) Promoting private sector participation in water supply in municipalities (*pourashavas*) iii) Implementing arsenic mitigation measures in arsenic-affected villages iv) Supporting development of adequate regulations, monitoring, capacity building, and training v) Supporting the development of a local credit market for village piped water supply vi) Implementing a monitoring and evaluation system for the project
3	Sanitation, Hygiene, and Water Supply Project (GOB-UNICEF)	January 2006 to December 2010	GOB: 1,175.92 PA: Tk 4,076.90 million	i) Reduce mortality, morbidity, and malnutrition due to water- and excreta-related diseases, especially among poor woman and children ii) Improve standard of hygiene behaviors on a sustainable basis iii) Improve access to safe water in unserved and underserved areas, including those suffering from arsenic contamination iv) Increase sanitation coverage to 100% in program areas as per GOB goal by 2010
4	Water Supply and Sanitation in Coastal Belt Project (GOB-DANIDA)	January 2006 to June 2008	350 million	i) Improve hygiene behaviors/practices ii) Promote community-led total sanitation iii) Increase the coverage of safe water supply services
5	Secondary Towns Water Supply and Sanitation Sector Project (GOB-ADB)	August 2006 to June 2012	GOB: 1,430.97 PA: 3,425.00 RPA: Tk 319.764 million	i) Increase the water supply coverage up to 90% from the present coverage of 29% (average)by the year 2015 with an additional inclusion of 1,948,000 people under the water supply system ii) Increase the sanitation coverage from 74% to 100% by the year 2010, fulfilling the GOB commitment "Sanitation for all by 2010" iii) Improved capacity of secondary towns to plan, implement, operate, manage, maintain, and finance water supply and sanitation investments iv) Improve capacity of Department of Public Health and Engineering (DPHE)to plan, design, supervise, monitor, and provide technical assistance to local water utilities and sanitation units

table continues next page

Sl.	Title	Period	Cost (Tk million)	Objective
6	Bangladesh Environmental Technology Verification-Support to Arsenic Mitigation (BETV-SAM)	July 2005 to June 2009	GOB: 5.78 PA: Tk 674.22 million	i) Test and verify the performance of 7 to 10 new technologies "screened in" for testing and verification ii) Ensure that the technologies verified in Environmental Technology Verification-Arsenic Mitigation (ETV-AM) and under this project stand the "test of time" in sustained use in real field conditions (monitoring): establish which arsenic removal technologies are fiscally and socially viable (fiscal and social monitoring and evaluation). iii) Secure the institutional sustainability of arsenic technology verification in Bangladesh iv) Provide early support to the GOB and private sector actors in the development of plans and programs to ensure that ETV-AM technologies eventually developed on a large-scale basis in Bangladesh are properly maintained and replaced when required
7	Repair, Rehabilitation, and Development of Water Supply System in *Pourashavas*, including regeneration of Production Tubewells	July 1997 to June 1998	GOB: Tk 486.00 million	Repair, rehabilitation, and development of the existing water supply system in *pourashava* areas: i) Through regeneration of production tubewells that have experienced reduction of discharge ii) Ensure the supply of water to the target population iii) Reduce wastage of water iv) Improve pressure in water supply system network v) Make the water supply system sustainable through creating scope for increased service connection vi) Address adequately the locations having arsenic and iron pollution in water
8	Accelerated Development Programme for Water Supply and Sanitation in Chittagong Hill Tracts Districts	July 1998 to June 2008	GOB: Tk 382.5 million	i) Expansion and development of water supply and sanitation facilities in rural and urban areas of hill districts to improve health status of people by improving hygienic environment, by reducing the waterborne and fecal-related diseases, and by ensuring safe water use for all domestic purposes ii) Establish piped water supply and limited sanitation systems in upazila HQs of hill districts iii) Create awareness among the people and establish sustainable water supply and sanitation systems iv) Conduct intensive surveys, investigations, and research and development activities to establish a suitable water supply system in the hydrogeologically difficult areas of the hilly region v) Develop skilled workforce in the water supply and sanitation sector through an adequate training program vi) Establish a permanent setup for regular testing of water quality

table continues next page

SI.	Title	Period	Cost (Tk million)	Objective
9	Environmental Sanitation and Water supply with piped network in Thana Sadar and Growth Centre *Pourashavas* (first Phase)	July 2000 to June 2008	GOB: Tk 2,408.012 million	i) Ensuring availability of safe water to the people of project areas ii) Reduction of diarrhea and other waterborne diseases with provision of safe water iii) Improving living standards of the people with the safe water supply and waste disposal iv) Adopting measures including training for achieving sustainable water supply and waste management v) Developing skilled workforce in the water supply and sanitation sector through an adequate training program vi) Improving capacity of DPHE in hydrogeological investigation and mapping
10	18 District Towns Water Supply, Sanitation, and Drainage Project (Phase-II)	July 2000 to June 2008	GOB: Tk 305.71 million	i) Expansion and rehabilitation of water supply and sanitation infrastructures ii) Awareness building toward sustainable water supply and sanitation management iii) Establish laboratory for expansion of water testing facilities
11	Rajshahi City Water Supply (2nd phase) Project RDPP-1	July 2002 to June 2008	GOB: Tk 461.077 million	i) Ensure safe, adequate, and convenient water supply for Rajshahi city by renovation and development of water supply system ii) Make the water supply system self-sustained by improvement of management and institutional development
12	Rural water supply project throughout the country (5th phase)	July 2004 to June 2010	GOB: Tk 3,859.889 million	i) Reduce the incidence of diarrheal and other waterborne diseases by supplying safe water to rural population ii) Increase the number of tubewells/water points to increase use of safe water for all domestic purposes iii) Ensure community participation in the operation and maintenance of rural water supply iv) Increase water supply coverage in rural areas according to the "National Water and Sanitation Policy, 1998" v) Maintain water supply facilities in the rural areas during and after natural calamities
13	Water Supply and Environmental Sanitation Project in Mongla Pourashava	July 2004 to June 2008	GOB: Tk 171.463 million	i) Ensure safe water in the project area ii) Reduce the incidence of diarrheal and other waterborne diseases by supplying safe water iii) Ensure public living standards as well as a total environmental sanitation situation through supply of safe water and proper sanitation iv) Provide training and other related activities to obtain sustainable water supply and sanitation system in the project area
14	Water Supply, Sanitation and Drainage Project in Sylhet and Barisal City	July 2005 to June 2015	GOB: Tk 2,824.111 million	i) Ensure safe and adequate water supply in the cities by the development and renovation of water supply network ii) Expand and develop drainage systems of the cities iii) Develop solid waste management of the cities iv) Improve capacity of DPHE to plan, design, supervise, monitor, and provide technical assistance to local water utilities and sanitation units

table continues next page

Sl.	Title	Period	Cost (Tk million)	Objective
15	Rural Water Supply in South Western part of Bangladesh	July 2007 to June 2012	GOB: Tk 397.405 million	i) Investigate hydrogeological and hydrological conditions for identification of viable water sources ii) Provide arsenic safe and saline options on emergency basis in the highly affected areas iii) Increase the coverage of safe water supply services iv) Dissemination of appropriate water supply technologies through demonstration and motivation of people to build their own safe water points
16	National Sanitation Project (second Phase)	July 2007 to June 2011	GOB: Tk 1,000 million	i) Ensure sustainable sanitation facilities to the communities in residential and public places such as, residential areas, growth centers, bazaars, and by the side of railways, waterways, highways, and so on. ii) Improve awareness level of personal hygiene behaviors of the people iii) Achieve one of the MDG targets—100% sanitation for all by the year 2010 iv) Make low-cost sanitary latrine components available at upazila and union levels v) Involve local government institutions (LGIs) for sustainable sanitary environment to the local habitat vi) Develop suitable technological options in chars, above flood levels, in natural disaster periods, roadside, rail, and waterways through research and development
17	Bangladesh Rural Water Supply and Sanitation Project (BRWSSP)	May 2012 to June 2017	SDR 48.4 million	(i) Increase provision of safe water supply and hygienic sanitation in the rural areas of Bangladesh, where shallow aquifers are highly contaminated by arsenic and other pollutants such as salinity, iron, and bacterial pathogens (ii) Facilitate early emergency response.
18	Dhaka Water Supply and Sanitation Project (BDWSSP)	March 2009 to December 2015	SDR 50.97 million	Improve storm-water drainage in select catchments in Dhaka and improve DWASA's planning capacity
19	Chittagong Water Supply Improvement and Sanitation Project (CWSSP)	October 2010 to December 2015	SDR 155.52 million	Increase sustainable access to safe water and improved sanitation; support the establishment of a long-term water supply, sanitation, and drainage infrastructure development and operational management program in Chittagong
20	Water Management Improvement Project	November 2007 to June 2015	SDR 96.42 million	Improve water resources management by improving infrastructure and institutions through rehabilitating damaged water infrastructure, piloting the role of local communities, and enhancing the institutional performance of the country's principal water institutions, particularly the Bangladesh Water Development Board (BWDB) and Water Resources Planning Organization (WARPO)

Note: ADB = Asian Development Bank; DANIDA = Danish International Development Agency; IDB = Islamic Development Bank; GOB = Government of Bangladesh; HQ = Headquarter; MDG = Millennium Development Goal; PA = Project Aid; RPA = Reimbursable Project Aid; UNICEF = United Nations Children's Fund.

Bibliography

Ahmed, T., Mahfuz, M., Ireen, S., Ahmed, A. M. S., Rahman, S., Islam, M. M., Alam, N., Hossain, M. I., Rahman, S. M. M., Ali, M. M., Choudhury, F. P., and Cravioto, A. 2012. "Nutrition of Women and Children in Bangladesh: Trends and Directions for the Future." *Journal of Health Population and Nutrition* 30 (1): 1–11.

Bartram, J., and Cairncross, S., 2010. "Hygiene, Sanitation, and Water: Forgotten Foundations of Health." *PLOS Medicine* 7 (11): e1000367, November.

Bhutta, Z. A., Ahmed, T., Black, R. E., Cousens, S., Dewey, K., Giugliani, E., Haider, B. A., Kirkwood, B., Morris, S. S., Sachdev, H. P. S., and Shekhar, M. 2008. "What Works? Interventions for Maternal and Child Undernutrition and Survival." *The Lancet* 371: 417–40.

Black, R. E., Brown, K. H., and Becker, S. 1984. "Effects of Diarrhea Associated with Specific Enteropathogens on the Growth of Children in Rural Bangladesh." *Pediatrics* 73: 799–805.

Brown, K. H. 2003. "Diarrhea and Malnutrition." Presented as part of the symposium: Nutrition and Infection, Prologue and Progress Since 1968. *The Journal of Nutrition*, American Society of Nutritional Sciences.

Brown, K. H., Black, R. E., Lopez de Romana, G., and Kanashiro, H. C. 1989. "Infant-Feeding Practices and Their Relationship with Diarrheal and Other Diseases in Huascar (Lima), Peru." *Pediatrics* 83: 31–40.

Brown, K. H., Black, R. E., Robertson, A. D., and Becker, S. 1985. "Effects of Season and Illness on the Dietary Intake of Weaning during Longitudinal Studies in Rural Bangladesh." *American Journal of Clinical Nutrition* 1 (41): 343–55.

Cairncross, S., Bartram, J., Cumming, O. L., and Broklehurst, C. 2010. "Hygiene, Sanitation, and Water: What Needs to Be Done?" *PLOS Medicine* 7 (11): 1366, November.

Campbell, D. I., Elia, M., and Lunn, P. G. 2003. "Growth Faltering in Rural Gambia is Associated with Small Intestinal Barrier Function Leading to Endotoxemia and Systemic Inflammation." *Journal of Nutrition* 133: 1332–38.

Clasen, T., Boisson, S., Routray, P., Torondel, B., Bell, M., Cumming, O., Ensink, J., Freeman, M., Jenkins, M., Odagiri, M., Ray, S., Sinha, A., Suar, M., and Schimdt, W. -P. 2014. "Effectiveness of a Rural Sanitation Programme on Diarrhoea, Soil-Transmitted Helminth Infection, and Child Malnutrition in Odisha, India: A Cluster-Randomised Trial." *Lancet Glob Health* 2 (11): e645–53.

Curtis, V., Cairncross, S., Yonlim R. 2000. "Domestic Hygiene and Diarrhea: Pinpointing the Problem." *Tropical Medicine and International Health* 5: 22–32.

Dangour, A. D., Watson, L., Cummin, O., Boisson, S., Che, Y., Velleman, Y., Cavill, S., Allen, E., and Uauy, R. 2013. "Interventions to Improve Water Quality and Supply, Sanitation and Hygiene Practices, and Their Effects on the Nutritional Status of Children (Review)." *The Cochrane Collaboration*, John Wiley & Sons Ltd.

Engle, P., Menon, P., and Haddad. L. 1999. "Care and Nutrition: Concepts and Measurement" .*World Development* 27 (8): 1309–37.

Food and Agriculture Organization (FAO), World Food Programme (WFP), and International Fund for Agricultural Development (IFAD). 2012. *The State of Food Insecurity in the World 2012. Economic Growth is Necessary but Not Sufficient to Accelerate Reduction of Hunger and Malnutrition*. Rome: FAO.

Gillespie, S., Haddad, L., Mannar, V., Menon, P., Nisbett, N., and the Maternal and Child Nutrition Study Group. 2013. "The Politics of Reducing Malnutrition: Building Commitment and Accelerating Progress." *The Lancet* 382: 552–69.

Guerrant, R. L., Schorling, J. B., McAuliffe, J. F., and de Souza, M. A. 1992. "Diarrhea as a Cause and an Effect of Malnutrition: Diarrhea Prevents Catch-Up Growth and Malnutrition Increases Diarrhea Frequency and Duration." *American Journal of Tropical Medicine and Hygiene* 47: 28–35.

Haddat, L., Alderman, H., Appleton, S., Song, L., and Yohannes, Y. 2002. "Reducing Child Undernutrition: How Far Does Income Growth Take Us?" FCND Discussion Paper # 137, International Food Policy Research Institute, Washington, DC.

Hanchett, S., Krieger, L., Khan, M. H., Kullmann, C., and Ahmed, R. 2011. "Long-Term Sustainability of Improved Sanitation in Rural Bangladesh." Water and Sanitation Program: Technical Paper. World Bank, Washington, DC.

Heaver, R. 2005. *Strengthening Country Commitment to Human Development—Lessons from Nutrition*. Washington, DC: World Bank, .

Helen Keller International (HKI). 2006. "Bangladesh in Facts and Figures: 2005 Annual Report of the Nutritional Surveillance Project." HKI, Dhaka.

Humphrey, J. H. 2009. "Child Undernutrition, Tropical Enteropathy, Toilets and Handwashing." *The Lancet* 374: 1032–35.

icddr,b, UNICEF Bangladesh, Global Alliance for Improved Nutrition (GAIN), and the Institute of Public Health and Nutrition (IPHN). 2013. "National Micronutrients Status Survey 2011–12." Dhaka, Bangladesh.

icddr,b, WaterAid Bangladesh, Local Government Division. 2014. "Bangladesh National Hygiene Baseline Survey." Preliminary Report. Dhaka, Bangladesh.

Joint Monitoring Programme (JMP). 2009. "Datasheet of Sanitation and Drinking Water Coverage in Bangladesh." United Nations Statistics Division (UNSD) and United Nations Economic and Social Commission for Asia and the Pacific (ESCAP) Millennium Development Goals Meeting, Bangkok, January 14–16.

Kim, J. Y. 2013. "Time for Even Greater Ambition in Global Health." *The Lancet* 382 (9908), Online version, December 3.

Kramer, M. S., Chalmers, B., Hodnett, E. D., Sevkovskaya, Z., Dzikovich, I., Shapiro, S., Collet, J. P., Vanilovich, I., Mezen, I., Ducruet, T., Shishiko, G., Zubovich, V., Mknuik, D., Gluchanina, E., Dombrovskiy, V., Ustinocvitch, A., Kot, T., Bogdanovich, N., Ovchinikova, L., Helsing, E., and PROBIT Study Group (Promotion of Breastfeeding Intervention Trial). 2001. "Promotion of Breastfeeding Intervention Trial (PROBIT): A

Randomized Trial in the Republic of Belarus." *Journal of American Medical Association* 285: 413–12.

Levinson, F. J., and Balarajan, A. 2013. "Addressing Malnutrition Multisectorally: What Have We Learnt from Recent International Experience?" United Nations, August.

Levinson, J. F. 1999. "Searching for a Home: The Institutionalisation Issue in International Nutrition." Background paper for the World Bank-UNICEF nutrition assessment. World Bank, Washington, DC.

Levinson, J. F., and McLachlan, M. 1999. "How Did We Get There? A History of International Nutrition." In T. Marchione, *Scaling up/Scaling Down*. Amsterdam: Overseas Publishers Association.

Lin, A., Benjamin, F., Arnold, B.F., Afreen, S., Goto, R., Huda, T. M. N., Haque, R., Raqib, R., Unicomb, L., Ahmed, T., Coldford Jr., J. M., Luby, S. P. 2013. "Household Environmental Conditions Are Associated with Enteropathy and Impaired Growth in Rural Bangladesh." *American Society of Tropical Medicine and Hygiene* 89 (1): 130–37

Local Government Division. 1998. "National Policy for Safe Water and Sanitation 1998." Ministry of Local Government, Rural Development and Cooperatives, Government of the People's Republic of Bangladesh.

———. 2005a. "National Sanitation Strategy." Ministry of Local Government, Rural Development and Cooperatives, People's Republic of Bangladesh, March.

———. 2005b. "Pro Poor Strategy for Water Supply and Sanitation in Bangladesh", Ministry of Local Government, Rural Development and Cooperatives, Government of the People's Republic of Bangladesh, February.

———. 2011a. "National Strategy for Water Supply and Sanitation Hard to Reach Areas of Bangladesh." Ministry of Local Government, Rural Development and Cooperatives, Government of the People's Republic of Bangladesh, December.

———. 2011b. "Sector Development Plan (2011–25) Water Supply and Sanitation Sector in Bangladesh." Ministry of Local Government, Rural Development and Cooperatives, Government of the People's Republic of Bangladesh, November.

———. 2012. "National Hygiene Promotion Strategy for Water Supply and Sanitation Sector in Bangladesh 2012." Ministry of Local Government, Rural Development and Cooperatives, Government of the People's Republic of Bangladesh.

———. 2014a. "Bangladesh National Hygiene Baseline Survey." Ministry of Local Government, Rural Development and Cooperatives, Government of the People's Republic of Bangladesh, Preliminary Report, June.

———. 2014b. "National Strategy for Water Supply and Sanitation." Ministry of Local Government, Rural Development and Cooperatives, Government of the People's Republic of Bangladesh, Draft Final, August 31.

Mara, D., Lane, J. Scott, B., and Trouba, D. 2010. "Sanitation and Health." *PLOS Medicine* 7 (11), November.

Martorell, R., Yarbrough, C., and Klein, R. E. 1980. "The Impact of Ordinary Illnesses on the Dietary Intakes of Malnourished Children." *American Journal of Nutritional Sciences* 33: 345–50.

Maxwell, S., and Conway, T. 2000. "New Approaches to Planning." Operations Evaluation Department Working Paper Series No. 14. World Bank, Washington, DC.

Mbuya, N., and Ahsan, K. Z. A. 2013. "The National Nutrition Program (NNP) Survey, 2010: Better, But Not Enough—An Assessment of Area Based Community Nutrition (ABCN) Services in Bangladesh." Report No: ACS3203, SASHN, World Bank, Washington, DC.

Ministry of Water Resources. 1999. "National Water Policy." Government of the People's Republic of Bangladesh, January 30.

National Institute of Population Research and Training (NIPORT). 2013."Utilization of Essential Services Delivery Survey 2013 (UESD)." Provisional Report. Dhaka, Bangladesh.

National Institute of Population Research and Training (NIPORT), Mitra and Associates, and MEASURE DHS. Various years. "Bangladesh Demographic and Health Survey 2007." Dhaka, Bangladesh.

———. 2013. "Bangladesh Demographic and Health Survey 2011." January. Dhaka, Bangladesh.

Newman, J., 2013. "How Stunting is Related to Having Adequate Food, Environmental Health and Care: Evidence from India, Bangladesh, and Peru." World Bank, Washington, DC.

Ngure, F. M., Humphrey, J., Mbuya, M. N. N., Majo, F., Mutasa, K., Govha, M., Mazarura, E., Chasekwa, B., Prendergast, A. J., Curtis, V., Boor, K. J., Stolzfus, R.F. 2013. "Formative Research on Hygiene Behaviors and Geophagy among Infants and Young Children and Implications of Exposure to Fecal Bacteria." *American Journal of Tropical Medicine and Hygiene* 89 (4): 709–716,

Ngure, F. M., Reid, B. M., Humphrey, J. H., Mduduzi, N. M., Pelto, G., Stoltzfus, R. J. 2014. "Water, Sanitation, and Hygiene (WASH), Environmental Enteropathy, Nutrition, and Early Child Development: Making the Links." *Annals of the New York Academy of Sciences* 1308 (2014): 118–28, 2014 New York Academy of Sciences.

Popkin, B. M., Adair, L., Akin, J. S., Black, R., Briscoe, J., and Flieger, W. 1990. "Breastfeeding and Diarrheal Morbidity." *Pediatrics* 86: 874–82.

Power and Participation Research Center (PPRC) and United Nations Development Programme (UNDP). 2012. "Ground Realities and Policy Challenges." *Social Safety Nets in Bangladesh*, Volume 2. Dhaka, Bangladesh.

Republic of Kenya, UNICEF, and World Bank. 2007. "Are Your Hands Clean Enough? Study Findings on Handwashing with Soap Behavior in Kenya." World Bank, Washington, DC.

Rokx, C. 2000. "Who Should Implement Nutrition Interventions? The Application of Institutional Economics to Nutrition and Significance of Various Constraints to the Implementation of Nutrition Interventions." HNP Discussion Series, World Bank, Washington, DC.

Rowland, M. G. M., Cole, T. J., and Whitehead, R. G. 1977. "A Quantitative Study into the Role of Infection in Determining Nutritional Status in Gambian Village Children." *British Journal of Nutrition* 37: 441–50.

Schmidt, C. W. 2014. "Beyond Malnutrition: The Role of Sanitation in Stunted Growth." *Environmental Health Perspectives* 122(11): A299–303.

Schmidt, W. -P., Aunger, R., Coombes, Y., Maina, P.M., Matiko, C. N., Biran, A., Curtis, V. 2009. "Determinants of Handwashing Practices in Kenya: The Role of Media

Exposure, Poverty and Infrastructure." *Tropical Medicine and International Health* 14 (12): 1534–41.

Scrimshaw, N. S., Taylor, C. E., and Gordon, J. E. 1968. *Interactions of Nutrition and Infection*. World Health Organization Monograph Series No. 57, World Health Organization, Geneva.

Smith, L. C., and Haddad, L. 2000. "Explaining Child Malnutrition in Developing Countries—A Cross Country Study." IFPRI Discussion Paper. International Food Policy Research Institute, Washington, DC.

Somanathan, A., and Mahmud, I. 2008. "Multisectoral Approaches to Addressing Malnutrition in Bangladesh: The Role of Agriculture and Microcredit." South Asia Sector for Human Development, World Bank, Washington, DC.

Subramanian, S., Huq, S., Yatsunenko, T., Haque, R., Mahfuz, M., Alam, M. A., Benezra, A., DeStefano, J., Meier, M. F., Muegge, B. D., Barratt, M. J., VanArendonk, L. G., Zhang, Q., Province, M. A., Petri Jr., W. A., Ahmed, T., and Gordon, J. I. 2014. "Persistent Gut Microbiota Immaturity in Malnourished Bangladeshi Children." Macmillan Publishers Limited, June.

The Institute of Public Health and Nutrition (IPHN). 2011. "Operational Plan of National Nutrition Services (2011–2016)." The Ministry of Health and Family Welfare, The Government of Bangladesh, December.

The Ministry of Health and Family Welfare. 2011. "National Health Policy 2011: Good Health is the Key to Development." Government of the People's Republic of Bangladesh.

United Nations Children's Fund (UNICEF). 1990. "Strategy for Improved Nutrition of Children and Women in Developing Countries." UNICEF Policy Review, UNICEF, New York.

———. 2006. "Global Framework for Action for Action." Revised Draft, UNICEF, New York, December.

———. 2013. "Improving Child Nutrition: The Achievable Imperative for Global Progress." UNICEF, New York, April.

United Nations. 2014. *The Millennium Development Goals 2014*. New York: United Nations.

Velleman, Y., and Pugh, I. 2013. "Undernutrition and Water and Sanitation and Hygiene." WaterAid and Share, UK.

Velleman, Y., Mason, E., Graham, W., Benova, L., Chopra, M., Campbell, O. M. R., Gordon, B., Wijesekera, S., Hounton, S., Mills, J. E., Curtis, V., Afsana, K., Boisson, S., Magoma, M., Cairncross, S., and Cumming, O. 2014. "From Joint Thinking to Joint Action: A Call to Action on Improving Water, Sanitation, and Hygiene for Maternal and Newborn Health." *PLOS Medicine* 11 (12): e1001771, December.

WASHplus Project. 2013. "Integrating Water, Sanitation, and Hygiene into Nutrition Programming." USAID, New York.

Water Resources Planning Organization (WARPO). 2001. "National Water Management Plan Development Strategy." Ministry of Water Resources, Government of the People's Republic of Bangladesh, June.

World Bank. 2006. "Agriculture and Achieving Millennium Development Goals." World Bank, Washington, DC.

————. 2011. "South Asia Regional Assistance Strategy for Nutrition (2011–2016)." World Bank, Washington, DC.

————. 2012. "Bangladesh: Towards Accelerated, Inclusive and Sustainable Growth—Opportunities and Challenges." Poverty Reduction and Economic Management Sector Unit, South Asia Region, World Bank, Washington, DC.

————. 2013a. "Bangladesh Poverty Assessment: Assessing a Decade of Progress in Reducing Poverty, 2000–2010." Bangladesh Development Series Paper No. 31, World Bank Office, Dhaka, June.

————. 2013b. "Making Sanitation Marketing Work: The Bangladesh Story." Water and Sanitation Program, World Bank, Washington, DC, December.

————. 2014. "What Progress in the Implementation of the National Nutrition Services (NNS) Program? Results of an Operational Assessment." Health, Nutrition and Population Global Practice, World Bank, Washington, DC.

World Bank, DFID, Government of Japan, and Rapid Social Response. 2013. "Improving Nutrition through Multisectoral Approaches." World Bank, Washington, DC, January.

World Health Organization, UNICEF, "Progress on Sanitation and Drinking-Water 2013", Joint Monitoring Programme for Water Supply and Sanitation (JMP).

World Health Organization, UNICEF, "Progress on Sanitation and Drinking-Water 2014", Joint Monitoring Programme for Water Supply and Sanitation (JMP)

World Health Organization (WHO). 2002. "Managing Water in the Home: Accelerated Health Gains from Improved Water Supply." Water, Sanitation and Health, Department of Protection of the Human Environment, WHO, Geneva.

Yael, V., Elizabeth, M., Wendy, G., Lenka, B., Mickey, C., Oona M. R. C., Bruce, G., Sanjay, W., Sennen, H., Joanna, E. M., Val, C., Kaosar, A., Sophie, B., Moke, M., Sandy, C., Oliver, C. 2014. "From Joint Thinking to Joint Action: A Call to Action on Improving Water, Sanitation, and Hygiene for Maternal and Newborn Health." PLOS Medicine 11 (12): e1001771, December 2014

Yusuf, H. K. M., Rahman, A. K. M. M., Chowdhury, F. P., Mohiduzzaman, M., Banu, C. P., Sattar, M. A., and Islam, M. N. 2008. "Iodine Deficiency Disorders in Bangladesh, 2004–05: Ten Years of Iodized Salt Intervention Brings Remarkable Achievement in Lowering Goiter and Iodine Deficiency among Children and Women." Asia Pacific Journal on Clinical Nutrition 17 (4): 620–28.

Environmental Benefits Statement

The World Bank is committed to reducing its environmental footprint. In support of this commitment, the Publishing and Knowledge Division leverages electronic publishing options and print-on-demand technology, which is located in regional hubs worldwide. Together, these initiatives enable print runs to be lowered and shipping distances decreased, resulting in reduced paper consumption, chemical use, greenhouse gas emissions, and waste.

The Publishing and Knowledge Division follows the recommended standards for paper use set by the Green Press Initiative. Whenever possible, books are printed on 50 percent to 100 percent postconsumer recycled paper, and at least 50 percent of the fiber in our book paper is either unbleached or bleached using Totally Chlorine Free (TCF), Processed Chlorine Free (PCF), or Enhanced Elemental Chlorine Free (EECF) processes.

More information about the Bank's environmental philosophy can be found at http://crinfo.worldbank.org/wbcrinfo/node/4.

green press
INITIATIVE